The Analytics of Risk Model Validation

Quantitative Finance Series

Aims and Objectives

- books based on the work of financial market practitioners, and academics
- presenting cutting edge research to the professional/practitioner market
- combining intellectual rigour and practical application
- covering the interaction between mathematical theory and financial practice
- to improve portfolio performance, risk management and trading book performance
- covering quantitative techniques

Market

Brokers/Traders; Actuaries; Consultants; Asset Managers; Fund Managers; Regulators; Central Bankers; Treasury Officials; Technical Analysts; and Academics for Masters in Finance and MBA market.

Series Titles

Return Distributions in Finance
Derivative Instruments: Theory, Valuation, Analysis
Managing Downside Risk in Financial Markets
Economics for Financial Markets
Performance Measurement in Finance
Real R&D Options
Advanced Trading Rules, Second Edition
Advances in Portfolio Construction and Implementation
Computational Finance
Linear Factor Models in Finance
Initial Public Offerings
Funds of Hedge Funds
Venture Capital in Europe
Forecasting Volatility in the Financial Markets, Third Edition
International Mergers and Acquisitions Activity Since 1990
Corporate Governance and Regulatory Impact on Mergers and Acquisitions
Forecasting Expected Returns in the Financial Markets
The Analytics of Risk Model Validation

Series Editor

Dr Stephen Satchell

Dr Satchell is a Reader in Financial Econometrics at Trinity College, Cambridge; visiting Professor at Birkbeck College, City University Business School and University of Technology, Sydney. He also works in a consultative capacity to many firms, and edits the *Journal of Derivatives and Hedge Funds*, *The Journal of Financial Forecasting*, *Journal of Risk Model Validation* and the *Journal of Asset Management*.

The Analytics of Risk Model Validation

Edited by

George Christodoulakis
Manchester Business School, University of Manchester, UK

Stephen Satchell
Trinity College, Cambridge, UK

AMSTERDAM • BOSTON • HEIDELBERG • LONDON • NEW YORK • OXFORD
PARIS • SAN DIEGO • SAN FRANCISCO • SINGAPORE • SYDNEY • TOKYO

Academic Press is an imprint of Elsevier

Academic Press is an imprint of Elsevier
30 Corporate Drive, Suite 400, Burlington, MA 01803, USA
84 Theobald's Road, London WC1X 8RR, UK
525 B Street, Suite 1900, San Diego, CA 92101-4495, USA

First edition 2008

British Library Cataloguing in Publication Data
A catalogue record for this book is available from the British Library

Library of Congress Cataloging-in-Publication Data
A catalog record for this book is available from the Library of Congress

ISBN: 978-0-7506-8158-2

For information on all Academic Press publications
visit our website at books.elsevier.com

Transferred to Digital Printing in 2010

Contents

About the editors

Dr George Christodoulakis is an expert in quantitative finance, focusing on financial theory and the econometrics of credit and market risk. His research work has been published in international refereed journals such as *Econometric Reviews*, the *European Journal of Operational Research* and the *Annals of Finance* and he is a frequent speaker at international conferences. Dr Christodoulakis has been a member of the faculty at Cass Business School City University and the University of Exeter, an Advisor to the Bank of Greece and is now appointed at Manchester Business School, University of Manchester. He holds two masters degrees and a doctorate from the University of London.

Dr Stephen Satchell is a Fellow of Trinity College, Reader in Financial Econometrics at the University of Cambridge and Visiting Professor at Birkbeck College, City University of Technology, at Sydney, Australia. He provides consultancy for a range of city institutions in the broad area of quantitative finance. He has published papers in many journals and has a particular interest for risk.

About the contributors

Sumit Agarwal is a financial economist in the research department at the Federal Reserve Bank of Chicago. His research interests include issues relating to household finance, as well as corporate finance, financial institutions and capital markets. His research has been published in such academic journals as the *Journal of Money, Credit and Banking*, *Journal of Financial Intermediation*, *Journal of Housing Economics* and *Real Estate Economics*. He has also edited a book titled *Household Credit Usage: Personal Debt and Mortgages* (with Ambrose, B.).

Prior to joining the Chicago Fed in July 2006, Agarwal was Senior Vice President and Credit Risk Management Executive in the Small Business Risk Solutions Group of Bank of America. He also served as an Adjunct Professor in the finance department at the George Washington University. Agarwal received a PhD from the University of Wisconsin-Milwaukee.

Joseph L. Breeden earned a PhD in physics in 1991 from the University of Illinois. His thesis work involved real-world applications of chaos theory and genetic algorithms. In the mid-1990s, he was a member of the Santa Fe Institute.

Dr Breeden has spent the past 12 years designing and deploying forecasting systems for retail loan portfolios. At Strategic Analytics, which he co-founded in 1999, Dr Breeden leads the design of advanced analytic solutions including the invention of Dual-time Dynamics. Dr Breeden has worked on portfolio forecasting, stress testing, economic capital and optimization in the US, Europe, South America and Southeast Asia both, during normal conditions and economic crises.

Souphala Chomsisengphet is Senior Financial Economist in the Risk Analysis Division at the Office of the Comptroller of the Currency (OCC), where she is responsible for evaluating national chartered banks' development and validation of credit risk models for underwriting, pricing, risk management and capital allocation. In addition, she conducts empirical research on consumer behavioral finance, financial institutions and risk management. Her recent publications include articles in the *Journal of Urban Economics*, *Journal of Housing Economics*, *Journal of Financial Intermediation*, *Real Estate Economics, and Journal of Credit Risk*.

Prior to joining the OCC, Chomsisengphet was an economist in the Office of Policy Analysis and Research at the Office of Federal Housing Enterprise Oversight (OFHEO). She earned a PhD in Economics from the University of Wisconsin-Milwaukee.

Kevin Dowd is currently Professor of Financial Risk Management at Nottingham University Business School, where he works in the Centre for Risk and Insurance Studies. His research interests are in financial, macro and monetary economics, political economy, financial risk management and, most recently, insurance and pensions. His most recent book *Measuring Market Risk* (second edition) was published by John Wiley in 2005.

Klaus Duellmann is Director in the research section of the Department of Banking and Financial Supervision in the central office of the Deutsche Bundesbank in Frankfurt. There, he performs research in economic capital models, in particular for credit risk, market risk and the interaction of risks. He has been a member of various working groups of the Basel Committee on Banking Supervision. He is Associate Editor of the *Journal of Risk Model Validation*. He holds a PhD from the faculty of business administration at the University of Mannheim, graduated in mathematics from the Technical University of Darmstadt and in business administration from the University in Hagen.

Wayne Holland is Senior Lecturer in the Operations group at Cass Business School, City University London, and Deputy Director for the upcoming Centre of Operational Excellence, London. He has a PhD in queueing analysis from Cardiff. His areas of interest lie in bootstrap simulation methods, risk analysis, and simulation modelling applied to operational risk and supply-chain risk.

Christoph Kessler is Executive Director and works in the Risk Management team at UBS Global Asset Management. His work concentrates on the analytics used in the bank's proprietary risk management system and the estimation process for the risk models. He joined the former Swiss Bank Corporation in 1988 as Risk Manager in the newly emerging Derivatives markets and later moved into the asset management area. His academic career includes a Diploma from the University of Freiburg, a PhD from the University of Bochum in Mathematics and post-doc work at the University of Hull, with majors in Mathematical Logic and in Stochastic Processes.

Chunlin Liu is Assistant Professor of Finance with College of Business Administration, University of Nevada. He teaches courses in bank management, investment and international finance. His current research interests include banking, consumer finance and capital markets. He has published in the *Journal of Money, Credit, and Banking, Journal of Financial Intermediation, Journal of International Money and Finance, Journal of International Financial Markets, Institutions & Money, International Review of Economics & Finance, Southern Economic Journal, Quarterly Review of Economics and Finance, Journal of Economics and Finance* and the *Asia-Pacific Financial Markets*. Prior to his career in academia, he worked in the banking industry as a financial economist. Chunlin Liu received his PhD in Finance from University of Rhode Island. He is also a CFA charterholder.

Vichett Oung is a postgraduate in Finance, Econometrics and Statistics. He graduated from the ENSIIE, French Engineering School of Information Technology, and received his Master of Science from Aston University, as well as two Masters of Arts in both Finance and Statistics from CNAM University. He started his career in 1995 as a Financial Economist at the Commission Bancaire, the French Banking Supervisor, where he managed the banking research unit and was much involved at the international level within the context of the Basel II project, as a member of the Research Task Force of the Basel Committee. He developed a specific interest and expertise in credit risk model validation.

After the completion of Basel II, he has moved in 2004 to the field of monetary and financial economics upon joining the Banque de France as Deputy Head of the Monetary Analysis and Statistics Division.

Günter Schwarz is Managing Director and the Global Head of the Risk Management team at UBS Global Asset Management, where he is in charge of coordinating risk management research and support, and in particular the proprietary risk management systems and models of UBS Global Asset Management. He began his career in 1990 at the then Swiss Bank Corporation, working in the area of asset management and risk analysis most of the time. His academic background is a Diploma and a PhD in Mathematics from the University of Freiburg, specializing in Stochastic Processes and Mathematical Statistics.

ManMohan S. Sodhi is Head of the Operations group at Cass Business School, City University London. He is also Director of the upcoming Centre of Operational Excellence, London that includes operational risk among its research themes. He has a PhD in Management Science from University of California, Los Angeles and after teaching at the University of Michigan Business School for two years, he worked for a decade in industry with consultancies including Accenture before coming to Cass in 2002. His current research interests are in risk management processes and modelling associated with operations.

Dirk Tasche joined Fitch Ratings as Senior Director in the Quantitative Financial Research (QFR) group. Dirk is based in London and will focus on group's efforts regarding credit portfolio risk and risk scoring models. Prior to joining Fitch, Dirk was a risk analyst in the banking and financial supervision department of Deutsche Bundesbank, Frankfurt am Main. He was mainly involved in the European Union-wide and national German legal implementation of the Basel II Internal Ratings Based Approach (IRBA). Additionally, he was charged with research on economic capital models and their implementation in financial institutions. Prior to Deutsche Bundesbank, Dirk worked in the credit risk management of HVB, Munich, and as a researcher at universities in Germany and Switzerland. He has published a number of papers on measurement of financial risk and capital allocation.

Wei Xia is Executive Consultant in the Risk and Capital group, PricewaterhouseCoopers LLP UK, responsible for cross-asset class derivative valuations and quantitative market risk and credit risk consulting. Wei is also a PhD candidate in Quantitative Finance at Birkbeck College, University of London and visiting lecturer at University of International Business and Economics, Beijing, China. He was a quantitative developer at Winton Capital Management responsible for designing and developing an in-house risk measurement and reporting system.

Preface

The immediate reason for the creation of this book has been the advent of Basel II. This has forced many institutions with loan portfolios into building risk models, and, as a consequence, a need has arisen to have these models validated both internally and externally. What is surprising is that there is very little written that could guide consultants in carrying out these validations. This book aims to fill that gap.

In creating the book, we have become aware that many of these validation issues have been around for a long time and that the need for this book probably predates Basel II. Of particular interest for investment banks and asset management companies are the problems associated with the quantitative risk management of ones own money and client money.

Clients in particular can become litigious, and one of the key questions that arise is whether the risk of the client portfolio has been properly measured. To assess whether this is so requires the validation of the portfolio risk model. This area is virtually non-existent but has some features in common with Basel I. Thus, it is considered good practice to consider back-testing, scenario analysis and the like. Purveyors of risk models claim to test their products themselves, but they rarely make their models available for external validation. This means that the asset manager needs to take responsibility for the exercise.

As editors, we were delighted that a number of young and prominent researchers in the field were happy to contribute to this volume. Likewise, we thank the publishers for their understanding, Anne Mason who managed the document harmoniously and the Bank of Greece whose support for risk management helped bring about the creation of this project.

1 Determinants of small business default[*]

Sumit Agarwal[†], Souphala Chomsisengphet[‡] and Chunlin Liu[¶]

Abstract

In this paper, we empirically validate the importance of owner and business credit risk characteristics in determining default behaviour of more than 31 000 small business loans by type and size. Our results indicate that both *owner*- and *firm*-specific characteristics are important predictors of overall small business default. However, owner characteristics are more important determinants of small business *loans* but not small business *lines*. We also differentiate between small and large business accounts. The results suggest that owner scores are better predictors of small firm default behaviours, whereas firm scores are better predictors of large firm default behaviour.

1. Introduction

In this chapter, we develop a small business default model to empirically validate the importance of *owner* and the *business* credit bureau scores while controlling for time to default, loan contract structure as well as macroeconomic and industry risk characteristics. In addition, several unique features associated with the dataset enable us to validate the importance of the owner and business credit bureau scores in predicting the small business default behaviour of (i) spot market loans versus credit lines and (ii) small businesses below $100 000 versus between $100 000 and $250 000.

Financial institutions regularly validate credit bureau scores for several reasons. First, bureau scores are generally built on static data, i.e. they do not account for the time to delinquency or default.[1] Second, bureau scores are built on national populations. However, in many instances, the target populations for the bureau scores are region-specific. This can cause deviation in the expected and actual performance of the scores. For example, customers of a certain region might be more sensitive to business cycles and so the scores in that region might behave quite differently during a recession. Third, the

* The authors thank Jim Papadonis for his support of this research project. We also thank seminar participants at the Office of the Comptroller of the Currency, Office of Federal Housing Enterprise Oversight, Brent Ambrose, Michael Carhill, John Driscoll, Ronel Elul, Tom Lutton, Larry Mielnicki, and Nick Souleles for helpful discussion and comments. We are grateful to Diana Andrade, Ron Kwolek, and Tim Murphy for their excellent research assistance. The views expressed in this research are those of the authors and do not represent the policies or positions of the Office of the Comptroller of the Currency, of any offices, agencies, or instrumentalities of the United States Government, or of the Federal Reserve Bank of Chicago.
† Federal Reserve Bank of Chicago, Chicago, IL
‡ Office of the Comptroller of the Currency, Washington, DC
¶ College of Business Administration, University of Nevada, Reno, NV

bureau scores may not differentiate between loan type (spot loans versus lines of credit) and loan size (below $100K and above $100K), i.e. they are designed as one-size-fits-all.

However, it is well documented that there are significant differences between bank spot loans (loans) and lines of credit (lines). For example, Strahan (1999) notes that firms utilize lines of credit to meet short-term liquidity needs, whereas spot loans primarily finance long-term investments. Agarwal *et al.* (2006) find that default performance of home equity loans and lines differ significantly. Hence, we assess whether there are any differences in the performance of small business loans and lines, and if so, what factors drive these differences?

Similarly, Berger *et al.* (2005) argue that credit availability, price and risk for small businesses with loan amounts below and above $100K differ in many respects. Specifically, they suggest that scored lending for loans under $100K will increase credit availability, pricing and loan risk; they attribute this to the rise in lending to 'marginal borrowers'. However, scored lending for loans between $100K and $250K will not substantially affect credit availability, lower pricing and lesser loan risk. This is attributed to the price reduction for the 'non-marginal borrowers'. Their results suggest that size does affect loan default risk.

Overall, our results indicate that a business owner's checking account balances, collateral type and credit scores are key determinants of small business default. However, there are significant differences in economic contributions of these risk factors on default by credit type (loans versus lines) and size (under $100K versus $100K–250K). We find that the effect of owner collateral is three times as much on default for small business loans than for lines. This result is consistent with Berger and Udell's (1995) argument that a line of credit (as opposed to loan) measures the strength of bank–borrower relationship, and as the bank–firm relationship matures, the role of collateral in small business lending becomes less important. Our results also show that the marginal impact of a 12-month increase in the age of the business on lowering the risk of a small business defaulting is 10.5% for lines of credit, but only 5.8% for loans. Moreover, a $1000 increase in the 6-month average checking account balance lowers the risk of default by 18.1% for lines of credit, but only 11.8% for loans. Finally, although both owner and firm credit scores significantly predict the risk of default, the marginal impacts on the types of credits differ considerably. The marginal impact of a 10-point improvement in the *owner* credit score on lowering the risk of defaults is 10.1% for lines, but only 6.3% for loans. A similar 10-point improvement in the *firm* credit score lowers the risk of default by 6.3% for small business loans, but only 5.2% for small business lines. These results are consistent with that of Agarwal *et al.* (2006).

Comparing small businesses under $100K (small) and those between $100K and $250K (large), we find that the marginal impact of a 10-point improvement in the owner credit score in lowering the risk of default is 13.6% for *small* firms, but only 8.1% for *large* firms. On the contrary, the marginal impact of a 10-point improvement in the firm credit score in lowering the risk of default is only 2.2% for *small* firms, but 6.1% for the *larger* size firms. Furthermore, a $1000 increase in the 6-month average checking account balance lowers the risk of default by 5.1% for *small* firms, but by 12.4% for *large* firms. These results suggest that smaller size firms behave more like consumer credits, whereas larger size firms behave more like commercial credits and so bank monitoring helps account performance. These results are consistent with that of Berger *et al.* (2005).

The rest of the chapter is organized as follows. Section 1.2 discusses the data, methodology and summary statistics. Section 1.3 presents the empirical results for small business defaults by type (Section 1.3.1) and size (Section 1.3.2). Section 4 provides concluding remarks.

2. Data, methodology and summary statistics

2.1. Data

The data employed in this study are rather unique. The loans and lines are from a single financial institution and are proprietary in nature. The panel dataset contains over 31 000 small business credits from January 2000 to August 2002.[2] The majority of the credits are issued to single-family owned small businesses with no formal financial records. Of the 31 303 credits, 11 044 (35.3%) are *loans* and 20 259 (64.7%) are *lines* and 25 431 (81.2%) are under $100K and 5872 (18.8%) are between $100K and $250K. The 90-day delinquency rate for our dataset of loans and lines are 1.6% and 0.9%, respectively. The delinquency rates for credits under $100K and between $100K and $250K are 1.5% and 0.92%, respectively. It is worth mentioning some of the other key variables of our dataset. First, our dataset is a loan-level as opposed to a firm-level dataset. More specifically, we do not have information of all the loans a firm might have with other banks. Second, because these are small dollar loans, the bank primarily underwrites them based on the owners' credit profile as opposed to the firms credit profile. However, the bank does obtain a firm-specific credit score from one of the credit bureaus (Experian).[3] The owner credit score ranges from 1 to 100 and a lower score is a better score, whereas the firm credit score ranges from 1 to 200 and a higher score is a better score.

2.2. Methodology

For the purpose of this study, we include all accounts that are open as of January 2000, and exclude accounts with a flag indicating that the loan is never active, closed due to fraud/death, bankruptcy and default.[4] Furthermore, we also exclude all accounts that were originated before 1995 to simplify the analysis on account age. We follow the performance of these accounts from January 2000 for the next 31 months (until August 2002) or until they default.

We use a proportional hazard model to estimate the conditional probability of a small business defaulting at time t, assuming the small business is current from inception up to time $t-1$. Let $D_{i,t}$ indicate whether an account i defaults in month t. For instance, the business could default in month 24, then $D_{i,t} = 0$ for the first 23 months and $D_{i,24} = 1$, and the rest of the observations will drop out of the sample. We define default as two cycles of being delinquent, as most accounts that are two cycles delinquent (i.e. 60 days past due) will default or declare bankruptcy. Furthermore, according to the SBRMS report, 57% of banks use the two cycles delinquent as their standard definition of default and another 23% use one cycle delinquent as their definition of default.[5]

The instantaneous probability of a small business i defaulting in month t can be written as follows:

$$D_{i,t} = h_0(t) \exp(\beta' X_i(t)) \tag{1.1}$$

where $h_0(t)$ is the baseline hazard function at time t (the hazard function for the mean individual i-th sample), we use age (number of months) of the account to capture 'seasoning'[6] as a proxy for this baseline. $X_i(t)$ is a vector of time-varying covariates; β is the vector of unknown regression parameters to be estimated; and $\exp(\beta'X_i(t))$ is the exponential distribution specification that allows us to interpret the coefficients on the vector of X as the proportional effect of each of the exogenous variables on the conditional probability of 'completing the spell', e.g. small business loan terminating.

The time-varying exogenous variables (known as covariates) that are crucial to a small business' decision to default can be classified into five main risk categories as follows:

$$\beta'X_{i,t} = \beta_1 Owner_{i,t-6} + \beta_2 Firm_{i,t-6} + \beta_3 LoanContract_{i,t}$$
$$+ \beta_4 Macro_{i,t-6} + \beta_5 Industry_{i,t-6} \tag{1.2}$$

where $Owner_{i,t-6}$ represents specific characteristics of the owner that may be important in the risk of a small business defaulting, including owner credit score, owner collateral and average checking account balance. $Firm_{i,t-6}$ represents firm-specific firm characteristics that may affect default risks of the firm, including credit score for the business, firm collateral and months in business.[7] Finally, $LoanContract_{i,t-6}$ captures loan amount, risk premium spreads and internally generated behaviour score for the loan. $Macro_{i,t-6}$ captures county unemployment rate as well as 9 state dummies.[8] $Industry_{i,t-6}$ captures 98 two-digit SIC dummies.[9] Time-varying values of owner, firm, loancontract, macro and industry risks are lagged 6 months before default because of concerns about endogeneity. For instance, owner credit score at default would have severely deteriorated. This would bias our results towards the owner risk score being highly significant (reverse causality). Similarly, we want to control for unemployment rate before default and at the time of default.[10] The above explanatory variables are defined in Table 1.1. In addition, we also consider the expected sign on each coefficient estimate in Table 1.1 and provide some intuitions below.

Owner risks

The use of owner's personal assets as collateral[11] to secure a business enhances the creditor's claims of new assets (see Berger and Udell, 1995). Owners using personal assets to secure the loans or lines are less likely to pursue unnecessary risky projects as there is more at stake; therefore, small businesses using owner collateral are less likely to default. Next, we control for the owner credit score. The higher the owner score, the riskier the business owner, i.e. higher the risk of default.[12] A 6-month average checking account balance captures the liquidity position of a business owner. We expect this owner characteristic to be inversely related to default.[13]

Firm risks

Like owner collateral, firm collateral merely alters the claims of the creditors (Berger and Udell, 1995). Hence, firm collateral is expected to have negative impact on default risks. Similarly, firms with higher credit score are expected to be less risky and, thus, are less likely to default. Finally, a non-linear estimation for months in business should capture

Table 1.1 Variables, definitions and expected signs in the event of default

Variable	Definition	Expected Sign
Owner risks		
Owner collateral	Dummy variable indicating owner-specific collateral (mortgage, personal savings, etc.)	−
Owner score$_{t-6}$	Quarterly updated score measuring owner credit risk characteristics – higher score high risk	+
Average 6 months checking account balance$_{t-6}$	Six-month average checking account balance updated monthly	−
Firm risks		
Firm collateral	Dummy variable indicating firm-specific collateral (receivables, cash, etc.)	−
Firm score$_{t-6}$	Quarterly updated score measuring firm credit risk characteristics – lower score high risk	−
Months in business	Months in business as reported by the credit bureau	+
Months in business (squared)		−
Loan contract		
Loan amount	Loan amount at origination	−
Interest rate spread$_{t-6}$	Interest rate – prime rate	+
Internal risk rating$_{t-6}$	Bank-derived risk rating for the loan	+
Macro and industry risks		
Unemployment rate$_{t-6}$	County unemployment rate	+

the aging process of any business, and we expect the default rate to rise up to a certain age and then drop thereafter, i.e. younger accounts have a higher probability of default.

Contract structure

Internal risk rating is a behavioural score based on the performance of the loan. The higher the behavioural score, the higher the risk of a small business defaulting. Loan amount determines the *ex post* risk characteristics of the owner and the business. A higher loan amount implies that both the business and/or the owner are lower risk, and thereby should reduce the risk of default. In other words, the bank perceives the borrower to be lower risk, and so, it is willing to provide a higher loan amount.

Macroeconomic risks

We expect that small businesses facing higher local unemployment rate are subject to higher risks of default.

Industry risks

Control for differing risk profile by SIC industry code.

Table 1.2 Summary statistics for small business accounts by type and size

Variables	Type		Size	
	Loans	Lines	Small	Large
Number of accounts	11 044	20 259	25 431	5,872
Share of total	35.3%	64.7%	81.2%	18.8%
Owner risks				
Owner collateral	0.33	0.02	0.35	0.08
Owner score$_{t-6}$	76	79	82	61
Average 6 months checking account Balance$_{t-6}$	$33 987	$31 156	$28 724	$57 059
Firm risks				
Firm collateral	0.47	0.40	0.44	0.64
Firm score$_{t-6}$	136	102	114	122
Months in business	135	109	116	145
Loan contract				
Loan amount	$103 818	$79 740	$65 420	$197 425
Loan interest rate	7.48	7.42	7.49	6.84
Internal risk rating$_{t-6}$	5.19	5.14	5.17	5.07
Macro and industry risks				
Unemployment Rate$_{t-6}$	5.25	5.22	5.23	5.22

2.3. Summary statistics

Table 1.2 provides summary statistics for some of the key variables. About 33% of the loans and 35% of the small firms have personal collateral, whereas lines and large firms have less than 10% personal collateral. Conversely, the lines and large firms have significant amount of firm collateral. Additionally, over 50% of the lines do not have any collateral. The loan amount is three times as much for the large businesses in comparison with the small businesses. Although not statistically significant, the internal credit ratings for the lines of credit and large businesses reflect lower risk in comparison with loans and small businesses.

3. Empirical results of small business default

We first estimate the baseline hazard, as discussed in Gross and Souleles (2002), using a semiparametric model to understand the default rate differences of same age accounts over calendar time and cohort by type and size segments. The semiparametric model estimation does not assume any parametric distribution of the survival times, making the method considerably more robust. The baseline survival curves for small business *loans* are statistically different than those for the *lines* (see Figure 1.1). The line sample exhibits a relatively higher survival rates (i.e. lower probability of default) with account age, but the loan sample exhibits a relatively lower survival rate (i.e. higher probability of default) with account age. Next, the baseline survival curves for small business credits between

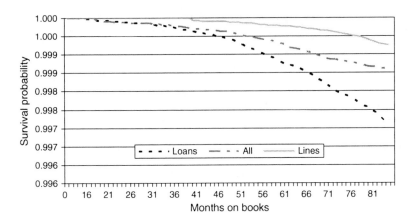

Figure 1.1 Survival curves for small business default by type

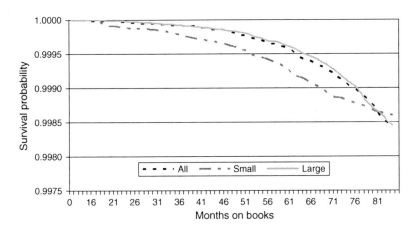

Figure 1.2 Survival curves for small business default by size

$100K and $250K are statistically different than those under $100K (see Figure 1.2). The larger credits exhibit a relatively higher survival rate (i.e. lower probability of default) with account age, but the smaller credits exhibit a relatively lower survival rate (i.e. higher probability of default) with account age.

Next, we estimate Equation 1.1 to assess the various factors that may impact the likelihood of a small business defaulting. We also conduct exhaustive robustness test by including quadratic specifications for the various risk variables, discrete dummies for some of the continuous variable, log transformations and others.

We first estimate the conditional probability of lines defaulting and loans defaulting separately. Table 1.3 summarizes the estimated impact of owner and firm risk on the likelihood of a small business defaulting, while controlling for loan contract structure and macroeconomic and industry risks. Below, we discuss how lines and loans do respond differently to their determinants, particularly owner- and firm-specific factors.

Table 1.3 Determinants of small business default – loans and lines

Variable	Type					
	Loans			Lines		
	Coefficient Value	Std. Error	t-Statistics	Coefficient Value	Std. Error	t-Statistics
Owner risks						
Owner collateral	−0.97823	0.35498	−2.76	−1.89872	1.24876	−1.52
Owner score$_{t-6}$	0.00103	0.00040	2.59	0.00299	0.00124	2.41
Average 6 months checking account balance$_{t-6}$	0.00000	0.00000	−2.30	−0.00001	0.00000	−3.06
Firm risks						
Firm collateral	−1.93484	0.33299	−5.81	−1.10893	0.33289	−3.33
Firm score$_{t-6}$	−0.00073	0.00033	−2.22	−0.00068	0.00023	−2.99
Months in business	0.00124	0.00340	0.36	0.04140	0.01239	3.34
Months in business (squared)	−0.00001	0.00000	−2.55	−0.00007	0.00002	−3.03
Loan contract						
Loan amount	0.00000	0.00000	−2.20	−0.00001	0.00000	−2.94
Risk premium$_{t-6}$	0.05283	0.01839	2.87	2.53459	0.33289	7.61
Internal risk rating$_{t-6}$	0.32349	0.04020	8.05	1.38989	0.13289	10.46
Macro and industry risks						
Unemployment rate$_{t-6}$	2.49890	0.73495	3.40	0.68933	0.56757	1.21
Quarter dummy	Yes			Yes		
SIC dummy	Yes			Yes		
State dummy	Yes			Yes		
Log likelihood	−627			−578		
number of observations	298 230			547 026		

3.1. Default behaviours of loans versus lines

Our results show that owner characteristics are less predictive of line defaults in comparison with loan defaults. The use of the owner's personal assets to secure loans, as opposed to lines, reduces the likelihood of loans defaulting. The finding that owner collateral is not a significant determinant of default for small business lines of credit is consistent with Berger and Udell (1995). Furthermore, a deterioration in the owner's as well as the firm's credit risk significantly raises the default risks of small businesses; however, the marginal impact varies between credit types. In Table 1.4, we show that the impact of a 10-point increase in the owner credit score (a deterioration of the credit risk of the owner) raises the default probability by 10.1% for loans, but only 6.3% for lines. On the contrary, a 10-point decline in the firm credit score (a deterioration of the credit risk of the firm) raises the default probability by 6.3% for loans, but only 5.2% for lines.

Moreover, we find that both owner and firm collateral are better predictor of default for loans than for lines. Owner collateral lowers the risk of default by 8.3% for loans, but only 2.9% for lines. Similarly, firm collateral lowers the risk of default by 4.4% for loans, but only 1.4% for lines.

Table 1.4 Marginal effects of owner and firm characteristics on small business default

Variables	Type		Size	
	Loans (%)	Lines (%)	Small (%)	Large (%)
Owner risks				
Owner collateral	− 8.3	−2.9	−2.2	−5.9
10 Point rise in owner score$_{t-6}$	10.1	6.3	13.6	8.1
$1000 Increase in average 6 months checking account balance$_{t-6}$	−11.8	−18.1	−5.1	−12.4
Firm risks				
Firm collateral	−4.4	−1.4	−0.7	−2.3
10 point drop in firm score$_{t-6}$	6.3	−5.2	2.2	6.1
12 Months rise of months in business	−5.8	−10.5	−7.9	−13.1

Equally important, the results show that the number of months in business is significantly positive, with the quadratic term significantly negative, for lines of credit. Small businesses that have been in business for an additional 1 year have a lower probability of default by 5.8% and 10.5%, respectively, for loans and lines. This result suggests that younger firms face higher risk of defaulting. However, the number of months in business is statistically insignificant in determining loan defaults. This would imply that even with age, loans are inherently more risky than lines.

The 6-month average checking account balance is highly significant in determining the default risks of lines and loans. However, the marginal impact of a $1000 rise in average checking account balance lowers the probability of default by 18% for lines, but only by 11% for loans. These results support the Mester *et al.* (forthcoming) argument that 'banks are special'.

3.2.   Default behaviours of small versus large credits

We investigate whether default behaviours of credits differ between small businesses with less than $100K (small) and those with debt between $100K and $250K (large). Table 1.5 summarizes the estimated coefficients of small business default for small and large debt accounts. These results are very interesting and provide evidence that small businesses under and over $100K have very different risk characteristics, as discussed below.

The risks of default between small businesses with credit amount of less than $100K and those with credit amount between $100K and $250K mainly differ in owner characteristics. For example, although both owner and firm collateral significantly reduce the likelihood of default, the impact is more striking for firms with credit amount between $100K and $250K (large) than for firms with credit amount less than $100K (small). Specifically, the use of owner collateral lowers the risk of default of large firms by 5.9%, but of small firms by only 2.2%. Similarly, the use of firm collateral lowers the risk of default of large firms by 2.3%, but of small firms by only 0.7%.

Furthermore, our results suggest that owner-specific score may be a better predictor of small firm default risks, whereas firm-specific score is a better predictor of large firm default behaviours. The reason lies in the magnitude of the marginal impact. For example, a 10-point increase in owner score (a deterioration in the owner's credit risk) raises the

Table 1.5 Determinants of small business default – small and large

Variable	Size					
	Small			Large		
	Coefficient Value	Std. Error	t-Statistics	Coefficient Value	Std. Error	t-Statistics
Owner risks						
Owner collateral	−0.33875	0.13489	−2.51	−8.32895	3.32478	−2.51
Owner score$_{t-6}$	0.00009	0.00004	2.31	0.23885	0.06924	3.45
Average 6 months checking account balance$_{t-6}$	−0.00001	0.00000	−2.12	−0.00001	0.00000	−2.83
Firm risks						
Firm collateral	−3.23899	0.44389	−7.30	−9.2381	6.7744	−1.36
Firm score$_{t-6}$	−0.00074	0.00034	−2.15	−0.00079	0.00039	−2.03
Months in business	0.00124	0.00798	0.16	0.02878	0.03848	0.75
Months in business (squared)	−0.00003	0.00001	−2.41	−0.00007	0.00003	−2.24
Loan contract						
Loan amount	−0.00001	0.00000	−4.67	−0.00001	0.00000	−3.50
Risk premium$_{t-6}$	0.04898	0.03289	1.49	0.33589	0.08327	4.03
Internal risk rating$_{t-6}$	0.54899	0.06325	8.68	0.73298	0.23775	3.08
Macro and industry risks						
Unemployment rate$_{t-6}$	0.13295	0.29835	0.45	0.03893	0.98355	0.04
Quarter dummy	Yes			Yes		
SIC dummy	Yes			Yes		
State dummy	Yes			Yes		
Log likelihood	−984			−591		
Number of observations	686 347			158 908		

probability of default by 13.6% for small credits and only by 8.1% large credits. On the contrary, a 10-point decline in firm score (a deterioration in the firm's credit risk) raises the probability of default by 2.2% for small credits, but by 6.1% for large credits. These results suggest that small credits behave more like consumer credits, whereas large credits behave more like commercial credits.

4. Conclusion

We empirically validate the importance of *owner* versus *firm* credit bureau score in determining default behaviours of small business loans, while controlling for time to default, the loan contract structure as well as macroeconomic and industry risks. We also compare and contrast the impact of *owner* and *firm* characteristics on small business default by type (loans versus lines) and size (under $100K versus $100K and $250K).

Our results indicate that both *owner*- and *firm*-specific characteristics are important predictors of overall business default. However, the economic impacts of owner characteristics significantly differ for small business *loans* than for *lines*. The marginal impact of owner collateral, owner credit risk improving and owner–bank relationship strengthening in lowering the risks of default is considerably larger for loans than for lines.

When we differentiate between small and large business accounts, our results suggest that the economic impact of owner and firm characteristics on small business default also differ considerably. For example, the marginal impact of an owner credit risk deteriorating on the probability of default is larger for small firms, whereas the marginal impact of a firm credit risk deteriorating on the probability of default is larger for large firm.

References

Agarwal, S., Ambrose, B.W., Chomsisengphet, S., *et al.* (2006). An empirical analysis of home equity loan and line performance. *Journal of Financial Intermediation*, 15(4), 444–469.

Berger, A.N., Frame, W.S. and Miller, N.H. (2005). Credit scoring and the availability, price, and risk of small business credit. *Journal of Money, Credit and Banking*, 37(2), 191–222.

Berger, A.N. and Udell, G.F. (1998). The economics of small business finance: The roles of private equity and debt markets in the financial growth cycles. *Journal of Banking and Finance*, 22, 613–73.

Gross, D.B. and Souleles, N.S. (2002). An empirical analysis of personal bankruptcy and delinquency. *Review of Financial Studies*, 15, 319–47.

Mester, L.J., Nakamura, L.I. and Renault M. (forthcoming). Checking accounts and bank monitoring. *Review of Financial Studies*.

Strahan, P.E. (1999). *Borrower Risk and the Price and Nonprice Terms of Bank Loans*. Federal Reserve Bank of New York Working Paper.

Notes

1. Credit bureau scores are built on historic data. The datasets are typically collected over a 12-month period with performance data for additional 24 months.
2. We very much liked to include a longer time period; unfortunately, data were not available. However, our data cover both the latter part of an economic growth and the early part of an economic recession. Moreover, as discussed in Berger and Udell (1998), an estimated 23.7% (52.7%) of small businesses disappear within the first 2 (4) years because of default/bankruptcy and other reasons. Hence, we should capture a sufficient number of defaults within the 2-year sample period (4 years in business because a business had to be operational for at least 18 months to qualify for a loan).
3. See, www.experian.com
4. As discussed in Gross and Souleles (2002), this "makes the data stationary." We conduct robustness tests by excluding accounts that default in the first 2 months since January 2000. The results are qualitatively the same.
5. For the purpose of this study, alternate definitions of default were also considered. Specifically, we defined default as 90 days past due and the results are robust to the definition of default.
6. Loan age "account seasoning" is modelled as a polynomial also discussed by Gross and Souleles (2002). The Cox Model does not explicitly report the coefficient values for the loan age variable, but the survival curves do provide the impact of account seasoning in small business default.
7. Months in business does not necessarily equal to loan age. As over 90% of the accounts in the dataset have been in business anywhere from 6 months to 24 months before applying for a loan at the financial institution.
8. Our data are primarily from the ten New England states. The dummies control for any states laws or state-specific macro economic polices.

9. For the purposes of this study, we interact the state and SIC dummy variables so that we have a state-specific control variable for the same SIC across states. This will help isolate any state-specific effect on a particular two-digit SIC code.
10. A related issue is the appropriate lag structure for these variables. As discussed in Gross and Souleles (2002), we could choose to lag these variables at account origination date, but that would not necessarily control for the risk composition between time of origination and the time since the beginning of the sample. Next, we could choose to control them at the time since the beginning of the sample or lag them 12 months from the time at default. We tried both these specifications, and the results are qualitatively the same.
11. There are over 70 distinct categories of collateral, but we have classified them into three broad categories, namely no collateral, owner collateral (residential mortgage, taxi medallions, stocks and bonds, bank deposits, gold/silver, etc.) and firm collateral (machinery, equipment, inventory, accounts receivable, letters of credit, etc.). This segmentation is consistent with Berger and Udell (1995, pp. 357) who describe owner and firm collateral as 'outside' and 'inside' collateral, respectively.
12. The score is counterintuitive, since traditional scores such as FICO lower risk with higher scores. This score is developed by Experian.
13. Mester, Nakamura and Renault (forthcoming) also conclude that checking account information does provide a 'relatively transparent window' in predicting small business credit deterioration. Evidence of a negative relationship between default and checking account balance.

2 Validation of stress testing models

*Joseph L. Breeden**

Abstract

Stress testing has gained importance in financial institutions with the introduction of Basel II. Although discussed from many perspectives, the predominant use for stress testing is in predicting how a portfolio would respond to changes in the macroeconomic environment. The future environment is encapsulated in a macroeconomic scenario for an extreme situation and then fed through a scenario-based forecasting model. Validating stress testing models is inherently difficult, because financial institutions do not have historical data representing portfolio performance through many severe recessions. Data availability presents challenges for standard in-sample/out-of-sample tests. This chapter discusses these limitations and describes a suite of tests that may be employed to determine the robustness of stress test models. Particular emphasis is given to retail portfolios, which have received very little attention in the literature.

1. Why stress test?

In many fields, stress testing is a routine part of the job. Architecture and engineering have made extensive use of stress testing for decades, with some very interesting case examples such as the Citibank Tower in Manhattan (Morgenstern, 1995). Although those fields have rich traditions and well-developed scientific literature on the subject, sadly very little of it is applicable to banking.

Just as a structural engineer would want to run simulations of extreme winds on a skyscraper, bankers need to simulate the impact of extreme macroeconomic environments on their portfolios. Therefore, a stress test model must contain explicit macroeconomic factors. All stress test models are scenario-based forecasts. The most comprehensive use of stress testing has been in tradable instruments, whereas the technology for stress testing the loan book has historically lagged far behind.

Basel II is increasing the visibility of stress testing by mandating stress testing for all of a bank's business lines. Basel II proscribes how banks will compute minimum capital requirements for their book of business. However, the calculation of a once-in-a-thousand-year event cannot be verified directly from the data of the implementing institution, because such events are not present in their historic data. To validate the capital calculation, paragraph 765 of the guidelines (Basel Committee on Banking Supervision, 2005) states clearly that stress testing will be employed to verify that the minimum capital computed under Basel II is sufficient to protect the bank against reasonably conservative

* President and Chief Operating Officer, Strategic Analytics Inc., Santa Fe, NM

scenarios of macroeconomic downturns. In the event of a shortfall, the minimum capital requirement will be increased to match the levels indicated by the stress test.

At first, the stress testing requirement of Basel II was largely ignored, but the last few years have seen this issue brought to the forefront. Central banks around the world have been releasing studies on best practices and proper use of stress testing for financial institutions. Documents released by the IMF (IMF, 2005), United Kingdom (Hoggarth and Whitley, 2003), Ireland (Kearns, 2004), Austria (Boss, 2002), the Czech Republic (Cihak, 2004) and Singapore (MAS, 2002) are just a few. Stress testing is also appearing in other contexts, such as a recent joint release by the US regulatory agencies on subprime lending, explicitly mandating stress testing for any institution engaged in such lending (OCC, 2004). This requirement is independent of whether those institutions are subject to Basel II.

Although most of these publications discuss best practices in creating stress testing models, none discuss how those models should be validated. Stress testing was enlisted as an indirect means of validating minimum capital calculations, but what validates the stress tests?

To discuss the validation of stress testing models, two potential areas of confusion must be addressed immediately. First, stress testing is not the same as sensitivity analysis. If we have a transition-matrix model predicting that 10% of BB-rated instruments are likely to be downgraded next year, asking what happens if that rises to 15% is not a stress test. It is a sensitivity analysis. Arbitrarily varying a model parameter will reveal how sensitive the portfolio is to that parameter, but says nothing about the likelihood of such a stress or how the portfolio might perform in an economic downturn.

The second confusion is that measuring goodness-of-fit when creating a model is not a validation. Portfolio forecasts and stress testing are extrapolation problems. The goal is to take the known time series and extrapolate to possible distant futures. With stress testing, we know in advance that we want to use our model to predict performance into environments never captured in the historical data – an inherently ill-posed problem.

Measuring goodness-of-fit while constructing the model is important, but verifying the model's robustness in extrapolation requires some form of out-of-time hold-out sample. This area of statistics is less well developed than interpolation models (which include credit rating and credit scoring models), where a large body of literature exists.

Stress testing model validation is largely concerned with verifying that the assumptions, structures and parameters incorporated into a model trained on historical data will persist far enough into the future and into extreme environments for the model to be useful.

2. Stress testing basics

Although this chapter does not have sufficient space for a full review of stress testing approaches, those models all have certain features in common by necessity. Those similarities lead the universal methods for validation.

The following is a list of features of stress test models:

- Use time series modelling.
- Incorporate macroeconomic data.
- Are not based upon the Basel II formulas.

Unlike scoring models or risk ratings, Bluhm *et al.* (2002) and Lando (2004) stress test models are explicitly time series models. Even when they incorporate risk ratings or scores, they must show how performance will evolve with time.

Time evolution of performance is assumed to be driven by macroeconomic data. The Basel II guidelines require that recession scenarios be considered, which implies that macroeconomic variables must be incorporated into those models. All of the validation work will assume that data exists and that we can build multiple models for comparison. To the extent that expert intuition is used in the model-building process, we will be testing the reproducibility of that intuition. If external models are being employed, many of these tests will not be possible, but the model provided should be asked to conduct appropriate validation tests.

Stress testing models will not use the minimum capital formulas described in the Basel II guidelines. Those formulas are meant to quantify the distribution of possible futures for the portfolio, parameterized by Probability of Default (PD) and the confidence level (99.9%). They do not predict the time evolution of the portfolio and do not incorporate any macroeconomic data. Therefore, the Basel II formulas cannot accept as input a serious recession to see how the portfolio would perform. In fact, the parameterization with PD is a crude method for adapting the distribution to the many types of products offered by financial institutions, which is probably one of the motivating factors behind the stress testing requirement.

In discussing stress testing models, we need to consider three asset classes: market tradable instruments, commercial loans and retail loans. These three classes usually require different modelling techniques to capture the unique dynamics of the products. Fortunately, even though the models are different, they have sufficiently similarity that we will be able to discuss common validation approaches.

2.1. Tradable instruments

All best-practices surveys find that stress testing continues to focus primarily on tradable instruments Commitee on the Global Financial System (CGFS, 2005), presumably because the ability to mark-to-market greatly simplifies the analysis for tradable instruments. Value-at-risk analysis (VaR) is a simulation approach designed to quantify the range of possible futures for the portfolio given the observed historic volatility. VaR is ubiquitous for analyzing tradable instruments, but as pointed out by the CGFS stress testing survey, it is not suitable for stress testing as it does not explicitly incorporate macroeconomic variables and does not consider extreme events outside the normal range of experience although variations have been proposed to achieve this (Kupiec, 1998).

Therefore, stress testing tradable instruments is done by creating econometric models relating changes in market value to changes in macroeconomic conditions. Given enough historical data to create such a relationship to macroeconomic drivers, scenarios for those variables are input to create a scenario-based forecast of the portfolio's future value.

2.2. Commercial lending models

Stress testing models for loan portfolios are considerably more challenging. In commercial lending, corporate risk ratings take the place of marking to market, so most models focus on transitions in credit ratings at the loan level.

A stress testing model would relate the probability of a rating downgrade to changes in macroeconomic conditions. These are essentially transition matrices with the probabilities conditioned on key macroeconomic drivers. The ratings transition model may also be conditioned on a number of other factors available to the lending institution, but the condition on macroeconomic conditions is the critical requirement. Given such a model, the stress test involves running a recession scenario and quantifying the impact upon the portfolio.

2.3. Retail lending models

Stress testing retail loan portfolios is arguably the most difficult. Consumers have credit scores, but those scores are explicitly intended not to change in any systematic way through macroeconomic cycles. Further, retail loans, especially subprime loans, have relatively high levels of predictable losses because of the consumer–product interaction lifecycle (Breeden, 2003).

With consumer loans, the loans with the highest predictable losses (subprime) have the lowest sensitivity to macroeconomic conditions. Conversely, the lowest risk loans (prime mortgage) have the highest sensitivity to macroeconomic conditions (Breeden, 2006). Portfolio managers usually say that subprime consumers are always in recession, whereas all prime losses are unexpected.

The result of these challenges is that retail lending stress test models are rarely successful unless they incorporate known factors such as vintage maturation. To capture the known structures, models must occur below the total portfolio level, such as the vintage or account level. A vintage in retail lending refers to a group of loans that were originated in the same period of time, such as an origination month. Performance for that vintage is then tracked as a function of time.

Experience has shown that when vintage effects are properly incorporated, stress test models tying performance residuals to macroeconomic drivers become possible. Without this adjustment, variation due to marketing plans and operational policies tends to destabilize econometric models (Figure 2.1).

2.4. Underlying similarities

Therefore, in considering validation approaches, we need to assume that a time series model has been built using macroeconomic variables as drivers to explain long-term portfolio performance as well as other predictive components. Generically, we can represent this as

$$\tilde{y}(a, t, v) = f\left(\mathbf{U}(v), \mathbf{W}(a), \mathbf{X}(t)\right) + \varepsilon(a, t, v), \tag{2.1}$$

where $\tilde{y}(a, t, v)$ is the forecast for portfolio performance metric $y(a, t, v)$, which could be PD, Loss Given Default (LGD), Exposure at Default (EAD), attrition, balance growth or other suitable portfolio measures. The factors $\mathbf{U}(v)$ capture instrument, loan or vintage-specific information such as scores or risk ratings, loan attributes, etc. $\mathbf{W}(a)$ are variables as a function of months-on-books, a, designed to capture the process known variously as seasoning, maturation or vintage lifecycle. $\mathbf{X}(t)$ are variables as a function of calendar date intended to capture macroeconomic impacts.

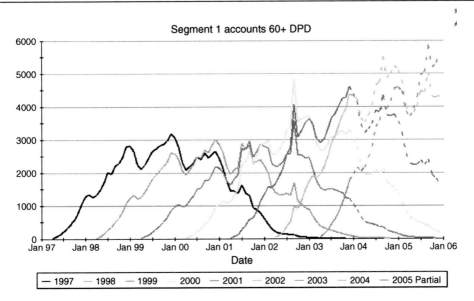

Figure 2.1 Delinquent accounts for a portfolio segment split by annual vintages (year of origination) showing strong maturation effects as well as seasonal and macroeconomic impacts

The model, f, will need to incorporate possible lagged impacts and other transforms of the input factors, using techniques such as transfer function models. In general, vector autoregressive models are typically constructed to incorporate the various impacts from outside factors. See Enders (2004), Wei (1989) and others for thorough discussions on how to create such models.

We assume that to create a stress test, a stressed scenario for the macroeconomic variables $\widetilde{X}(t)$ has been created and fed through the time series forecasting model f. Facility-level, loan-level and vintage-level forecasts will all need to be aggregated up to achieve total portfolio impacts.

To capture cross-correlations between different portfolio metrics or different segments, we need simply include the same factors $\widetilde{X}(t)$ in multiple models. Thus, as the scenario for the economy unfolds, the full impact can be felt in concert across the portfolio.

3. Overview of validation approaches

With these simple assumptions on the form of the model, we can ask a range of questions.

1. Is the model stable?
2. Does the model forecast well?
3. Are the results reproducible across segments?
4. Do we have reliable scenarios?

Stress test models need to forecast 1 year ahead. This is usually accomplished by modelling monthly performance data. Although tradable instruments may have longer histories, they are often in different macroeconomic regimes, meaning that only the last

10–20 years at most are applicable in the modelling. For retail lending, the longest data history in the United States is 14 years for mortgage defaults. Most institutions around the world currently have about 5 years of data for building and testing these models although the time series are expected to grow due to Basel II. Data sets for emerging markets are typically among the shortest.

Therefore, we can only expect to see one or two recessions in most data sets. This data limitation raises the greatest and least testable question of all.

1. Is a model created on recent macroeconomic stresses applicable to possible future environments?

Some of these questions are more amenable to testing than others. The following sections will explore the possibilities.

4.   Subsampling tests

For short time series, stability tests are the easiest to perform. If we split our data set into two or more samples, each with the same time range, and rebuild the stress test model independently on each part, are the models structurally equivalent? This subsampling is performed on a dimension other than time, because we want to compare models over the same time period.

4.1.   Random sampling

When building a stress test model, we assume the model applies to all the loans or facilities within that data set. If we randomly split the data set in half, we are assuming the same model would still apply to both pieces. When account level data is available, randomly sampling accounts seems the most obvious. For data aggregated to the vintage level, which is common in retail lending, splitting the data by alternating vintages is a natural approach.

When the subsamples have been created, the simple first step is to compute goodness-of-fit statistics for the overall model as applied to each of the subsets. As the model will still cover the same length of time for all subsets, the accuracy should be statistically equivalent on all subsets. This is, however, still just a comparison of in-sample model performance.

Testing the overall model on the subsets is not as good as asking whether those subsets would produce the same model. The best approach is to create new, independent models for each data subset to determine how structurally stable the results are. We expect the dependence of the model f on the macroeconomic factors $X(t)$ to be the same for all subsets. If the subsets select strongly for different dependencies, it would suggest that the modelling approach is not sufficiently robust.

The same techniques will apply to the loan- or vintage-specific factors $U(v)$. By creating functional relationships between performance and the factors $U(v)$ on different randomly generated data sets, the stability of the models can again be tested.

Lastly, we can use the sampling approach coupled with an out-of-sample test. By holding out the last year data, we can perform a weak comparison (because of the limited data length) of the models produced on each subset.

Several test statistics are available to measure whether the forecasts coming from two models are significantly different when applied to an out-of-sample period. Assume that we have two models, f_1 and f_2, built on two subsets of the in-sample portfolio of the time series. To compare the models, we test them on the same out-of-sample data set. The errors in their forecasts are e_1 and e_2, respectively. The following test statistics can then be applied.

F-Statistic

We begin by computing the mean-square-percentage error (MSPE) for each model over H out-of-sample time points.

$$MSPE = \frac{1}{H} \sum_{i=1}^{H} e_i^2 \tag{2.2}$$

By computing the ratio of the MSPE for the two models, placing the larger error in the numerator, we get the F statistic with H degrees of freedom.

$$F = \frac{\sum_{i=1}^{H} e_{1i}^2}{\sum_{i=1}^{H} e_{2i}^2} \tag{2.3}$$

If the F-statistic is significantly different from zero, then the models are not equivalent, and we have failed the stability test.

This test is only valid when the following three conditions are met:

1. Errors are zero mean and normally distributed.
2. Errors are serially uncorrelated.
3. Errors are not cross-correlated.

Granger–Newbold test

In many practical situations, the above conditions for applying the F-test cannot be met. Granger and Newbold (1976) developed an approach that also applies for series that do have cross-correlated errors. For their approach, they let $x_t = e_{1t} + e_{2t}$ and $z_t = e_{1t} - e_{2t}$. Then they compute

$$GN = \frac{r_{xz}}{\sqrt{(1 - r_{xz})/(H-1)}}, \tag{2.4}$$

where r_{xz} is the correlation between the two series x_t and z_t. The GN statistic has a t-distribution with degrees of freedom $H - 1$. If this measure is statistically different from zero, then the models are not equivalent in their accuracy, and the stability test fails.

Diebold–Mariano test

The GN test statistic uses a quadratic function of the errors to compare accuracy. This test has been generalized to any accuracy measure, a common choice being absolute error (Diebold and Mariano, 1995). Given an objective function $g(e)$ the mean modelling error is

$$\bar{d} = \frac{1}{H} \sum_{i=1}^{H} |g(e_{1i}) - g(e_{2i})| \qquad (2.5)$$

For models of equivalent accuracy, $\bar{d} = 0$. To test whether $\bar{d}$ is significantly different from zero, we compute

$$DM = \frac{\bar{d}}{\sqrt{(\gamma_0 + 2\gamma_1 + \ldots + 2\gamma_q)/(H + 1 - 2j + j(j-1)/H)}} \qquad (2.6)$$

If we let γ_i equal the ith autocovariance of the sequence $d_i = g(e_{1i}) - g(e_{2i})$, for models performing j step ahead forecasts, then the DM statistic is a t-distribution with $H - 1$ degrees of freedom. If the result is significantly different from zero, then the models are not equivalent.

For examples of the use of these tests and a discussion of their derivations, see Enders (2004). Keep in mind that these tests utilize a single out-of-sample data set for model comparison. That data set is probably not comprehensive relative to all possible future regimes in which the models will be applied. As such, these tests should be recognized as providing limited confidence about what can happen in the future.

4.2. Old versus new accounts

Random sampling is the best way to test the structural stability of the use of calendar-based factors, $X(t)$, such as macroeconomic impacts. However, we also want to verify that the maturation factors, $W(a)$, are also stable. By splitting accounts before and after an arbitrary origination date, we are testing that the accounts booked recently are similar in behaviour to those booked previously. One should assume that variations in credit quality will occur between the two groups, but correcting for minor variations in credit quality, we should still see the same dependence of PD, LGD and EAD on the number of months since origination.

Failure here is more common than failing the random sampling test. Failure occurs because the segment population is not stable over time. For example, if a retail lending business shifted its originations over the last few years from prime to include a significant subprime population, the expected maturation process for those recent originations would be significantly different from the prior years' bookings. Figure 2.2 shows an example of how the default rate maturation curve can change when stability fails.

This shift in default lifecycle would not be properly captured by an overall average maturation curve and would lead to time-varying estimation errors that would likely

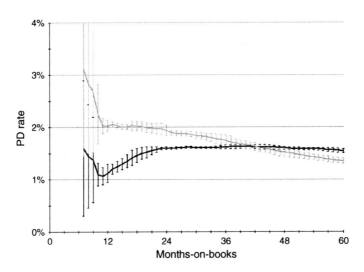

Figure 2.2 A comparison of two lines, one for each half of the vintages of a segment, showing that the dependence of their PD rates on months-on-books has shifted significantly over vintages. This segment fails the stability test

surface in the calibration to macroeconomic factors. To create a reliable stress test model, all of the model factors must be stable over the data set, not only the macroeconomic sensitivity. Although visually the stability failure in Figure 2.2 seems pretty clear, we can use the Granger–Newbold or Diebold–Mariano tests described in section 4.1 to test the hypothesis directly. If we create two series computing the differences between each of the curves in Figure 2.2 and the average curve obtained for the full data set,

$$e_{ji} = \mathbf{W}_j(a) - \overline{\mathbf{W}}(a),$$ (2.7)

where j represents the two series, i is the months-on-books index and the $\overline{\mathbf{W}}(a)$ are the maturation curves obtained by modelling the full data set. Then, either of the GN or DM tests can be used to see whether the maturation curves created for the two subsets are significantly different as compared to the full curve. Note that these errors are most likely correlated so the F-test of $F = \left(\sum_{i=1}^{H} e_{1i}^2\right) \Big/ \left(\sum_{i=1}^{H} e_{2i}^2\right)$ would be inappropriate.

Failing the split-vintage test suggests that the segmentation is insufficient to assure model stability. In the above example, subprime loans should not have been included in the modelling of prime loans. Over-segmentation is detrimental, because modelling error will rise as the event rate falls. For all data sets, a 'sweet spot' will exist in the trade-off between increased segmentation to improve estimation of maturation curves, transition matrices or whatever structural components might be needed versus the decreasing the segmentation to raise the event rate and diminish the estimation error.

5. Ideal scenario validation

Once we have verified the structural stability of the stress test model, the question any reviewer of that model will ask is, 'How good is it?' For a scenario-based model, this is a surprisingly difficult question.

In time series analysis, the typical response is to run an in-sample (in-time) and out-of-sample (out-of-time) validation. A model is built only on the in-sample data and then tested on the out-of-sample data.

The challenge with a pure out-of-sample test in scenario-based forecasting is that a scenario must be chosen for the out-of-sample period. Consequently, we will have difficulty distinguishing between a bad model and a bad scenario. Stress testing models will necessarily incorporate macroeconomic scenarios as input, as well as future marketing plans and account management changes. If the scenario is bad, no level of modelling sophistication will be sufficient to correct the forecast.

The solution to this problem is to run separate validation tests on the model and the scenario. To validate the model, we conduct an ideal scenario validation (ISV). Simply put, if we had used the 'ideal' scenario during the out-of-sample period, how accurate would the model have been. Our usual approach is to treat the estimation of maturation effects, $\mathbf{W}(a)$; vintage origination factors, $\mathbf{U}(v)$; seasonality, $\mathbf{X}_{SSN}(t)$ – a component of $\mathbf{X}(t)$ and the structural model combining these, f, as determinants of the model accuracy. They are estimated from the in-sample data. For the macroeconomic scenario, $\mathbf{X}_{ECON}(t)$ – components of $\mathbf{X}(t)$, we use the actual value of those factors during the out-of-sample period as the ideal scenario. Similarly, if any new originations are being included in the stress testing model, the actual bookings for those are used, because in theory, this is a known plan provided by marketing. Figure 2.3 shows the origin of the components used in the out-of-sample test.

When the test is run, the forecasting error is an indicator of the model accuracy. Admittedly, a 12-month test provides limited information, but as the model is validated across each segment and vintage, greater confidence is gained.

A high error rate for the ISV indicates either too much segmentation leading to high estimation errors or too little segmentation leading to specification errors in the factors or drift in the model structure. From a modelling perspective, the best approach is to vary the segmentation until an optimal value for the ISV is obtained. One straight-forward method is to over-segment the data, then aggregate structurally similar segments until the overall ISV has reached a minimum.

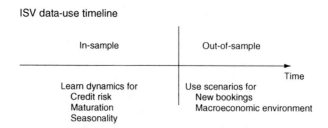

Figure 2.3 Diagram showing the proper use of in-sample and out-of-sample data for an ideal scenario validation

6. Scenario validation

A typical ISV might show a single-digit error rate over a 1-year horizon, but real-life use of the model should be expected to be less accurate. The problem is that scenario design is an inexact art, and macroeconomic forecasting is notoriously problematic. When one sales representative for a vendor of economic scenarios was asked how accurate they were, the response was 'It wouldn't make any sense to publish the accuracy of our scenarios, because we make a new one every quarter'.

Fortunately, the US Congressional Budget Office is a bit more forthcoming on their forecasting accuracy. They recently published a report (CBO, 2005) in which they analysed all of their 2-year forward forecasts made between 1982 and 2003, as well as those from the US Administration and the Blue Chip index of professional forecasters.

These errors swamp any ISV-measured modelling error. The report points out that all three forecasting groups are correlated in their errors, so there is little noise-reduction benefit in combining forecasts. They are most wrong at economic turning points, arguably when they are needed most. In addition, the forecasts are biased for many years following significant macroeconomic changes, such as the efficiency gains of the 1990s, which were not foreseen in their models.

This is all very bleak news for creating a baseline forecast using macroeconomic data considering that the typical business operational plan uses an 18-month horizon. Although the results in Table 2.1 clearly shows why we need the ISV approach for validating the underlying stress testing model, macroeconomic scenario accuracy is less critical to the use of those models for stress testing. The goal under Basel II is to conduct stress tests with extreme, improbable and yet plausible scenarios.

The challenge for stress testing scenarios is creating internal consistency. The scenario must be plausible in the sense that if unemployment rises dramatically, we cannot also assume that other macroeconomic indicators stay the same or improve. The Fair model (Fair, 1994) is one such publicly available model often used for generating self-consistent scenarios.

Aside from testing the internal consistency of a macroeconomic scenario, another challenge is to compute the probability of occurrence for a given macroeconomic scenario. Few, if any, macroeconomic forecasters can provide a quantified probability, but even if they could, it would not be useful for Basel II. If a recession scenario was assigned as one

Table 2.1 Mean percent error for 2-year-ahead forecasts from the congressional budget office, blue chip index, and US administration

Mean percent error by variable	CBO(%)	Blue chip(%)	Administration(%)
Growth rate for real output	130	109	141
Growth rate for nominal output	19	17	21
Inflation in the consumer price index	25	27	25
Nominal interest rate on 3-month treasury bills	28	29	28
Nominal long-term interest rate	10	10	12
Real interest rate on 3-month treasury bills	90	77	85

of the 1% worst recessions, it does not mean that it will produce a loss estimate in the 99th percentile of the distribution for a specific retail portfolio. For example, while commercial loans to the transportation sector may be very sensitive to an oil shock, retail loans will respond much more slowly and mostly through indirect impacts on employment and wages. The result is that we cannot know the severity of a scenario for retail lending just from an economist's ranking of that scenario according to GDP or some other measure. Thus, Basel II makes the vague request that a 'severe' recession be tested and compared against the minimum capital calculation, without specifically requesting a recession at the 99.9% severity level. Although regulators will probably put standards in place for scenario design, the difficulty of assigning probabilities to those scenarios will remain.

7.   Cross-segment validation

The greatest problem with time series analysis for retail portfolios is the limited data length. As mentioned earlier, the longest portfolio performance history in the United States is 14 years for mortgages, and most portfolios are much shorter. Worldwide, most practitioners have only one or two recessions at most to model against. Comparing regional economic differences within a portfolio is one way to augment this history.

   Rather than model an entire portfolio as one macroeconomic region, we can compare geographic regions that may have different macroeconomic conditions. For example, at a national level no housing price collapses have occurred in the United States in decades, but a regional or city-level segmentation will reveal housing price bubbles.

   With a geographic segmentation, the goal would be to create a stress test model applicable across all segments, but hold a few segments out during the model building so that they may be used as an out-of-sample test.

   Similar regional analyses may be possible for multinational portfolios in Europe or Asia if suitable calibrations are included for national differences in consumer behaviour.

8.   Back-casting

Every discussion of time series model validation needs to consider the issue of het-eroskedasticity. In this context, are recent recessions (i.e. volatility) representative of portfolio response to future recessions?

   One way to examine this problem is to look backward. By calibrating portfolio performance to macroeconomic factors, one can consider how the portfolio would have performed during previous macroeconomic environments. Financial institutions do not have internal data to verify that this back-cast performance was actually experienced, but the nature of the portfolio impact from those previous recessions can be compared to recent recessions. Was the rate of change in portfolio performance similar to what we observe today? Was the autocorrelation structure similar? At the most basic, does the model capture the impact of known prior recessions in a plausible manner?

   One unpublished study at a major multinational bank suggested that recessions since the early 1980s are similar in consumer loan response, but recessions prior to that began and ended more abruptly. This is interpreted as a sign of the impact of modern interest rate management by the Federal Reserve. If confirmed, the question becomes whether future recessions should be assumed to behave more like recent recessions or whether all eras should be considered when trying to create severe recessions. Such questions are more philosophical and political than the domain of model validation and thus will likely be resolved by the regulatory bodies.

9. Conclusions

Any stress testing model is necessarily a scenario-based forecasting model. To validate that model, one must distinguish between the accuracy of the predictable internal portfolio dynamics and the unpredictable or uncontrolled external impacts. We cannot have enough certainty in any macroeconomic scenario to get a highly accurate long-range portfolio forecast, but we can validate the predictable part of the model and feed it with plausible macroeconomic scenarios to measure the range of possible future portfolio performance.

References

Basel Committee on Banking Supervision (2005) *Basel II: International Convergence of Capital Measurement and Capital Standards: a Revised Framework*. Bank for International Settlements, November.

Bluhm, C., Overbeck, L. and Wagner, C. (2002) *An Introduction to Credit Risk Modeling*. Chapman & Hall/CRC, New York.

Boss, M. (2002) A macroeconomic credit risk model for stress testing the Austrian credit portfolio. *Financial Stability Report*. National Bank of Austria, December, pp. 62–82.

Breeden, J.L. (2003) Portfolio forecasting tools: what you need to know. *RMA Journal*, 78–87.

Breeden, J.L. (2006) Universal laws of retail economic capital. *RMA Journal*, 48–51.

Cihak, M. (2004) Designing stress tests for the Czech banking system. *International Research and Policy Note*, No. 3. Czech National Bank.

Committee on the Global Financial System (CFGS) (2005) *Stress Testing at Major Financial Institutions: Survey Results and Practice*, Report no. 24. Bank for International Settlements, January.

Congressional Budget Office (CBO) (2005) *CBO's Economic Forecasting Record: An Evaluation of the Economic Forecasts CBO Made from January 1976 Through January 2003*. The Congress of the United States, October.

Diebold, F. and Mariano, R. (1995) Comparing predictive accuracy. *Journal of Business and Economic Statistics*, 13, 253–63.

Enders, W. (2004) *Applied Econometric Time Series*. Wiley, USA.

Fair, R.C. (1994) *Testing Macroeconometric Models*. Harvard University Press, Cambridge, MA.

Hoggarth, G. and Whitley, J. (2003) Assessing the strength of UK banks through macroeconomic stress tests. *Financial Stability Review*. Bank of England, June, pp. 91–103.

Granger, C. and Newbold, P. (1976) Forecasting transformed series. *Journal Royal Statistical Society B*, 38, 189–203.

International Monetary Fund and the World Bank (IMF) (2005) *Implementation of Basel II—Implications for the World Bank and the IMF*, 22 July.

Kearns, A. (2004) Loan losses and the macroeconomy: a framework for stress testing credit institutions financial well-being. *Financial Stability Report*. Central Bank and Financial Services Authority of Ireland, pp. 111–121.

Kupiec, P. (1998) Stress-testing in a value at risk framework. *Journal of Derivatives*, 6(1), 7–24.

Lando, D. (2004) *Credit Risk Modeling: Theory and Applications*. Princeton University Press, Princeton.

Monetary Authority of Singapore (MAS) (2002) *Consultative Paper on Credit Stress-Testing*, 31 January.

Morgenstern, J. (1995) The fifty-nine story crisis. *The New Yorker*, 29 May, pp. 45–53.

Office of the Comptroller of the Currency (OCC), Board of Governors of the Federal Reserve System, Federal Deposit Insurance Corporation, and Office of Thrift Supervision (2004) *Expanded Guidance for Subprime Lending Programs*, January.

Wei, W. (1989) *Time Series Analysis: Univariate and Multivariate Methods*, Addison-Wesley.

3 The validity of credit risk model validation methods*

George Christodoulakis[†] and Stephen Satchell[‡]

Abstract

In this chapter, we discuss the nature, properties and pitfalls of a number of credit risk model validation methods. We focus on metrics of discriminatory power between sick and healthy loans, their association and their properties as random variables, which may lead to pitfalls in model validation processes. We conclude with a discussion of bootstrap and credit-rating combinations.

1. Introduction

The development of various types of credit risk models has its origins in the pricing of assets and has been further strengthened by the Basel Capital Accord, which allows for the determination of capital adequacy of credit institutions using internal rating models. This process has led the financial institutions as well as their supervisors to develop not only internal models but also validation methods to assess the quality and adequacy of those models. The need for credible assessment stems from the fact that using low-quality models could lead to sub-optimal allocation of capital as well as ineffective management of risks. The assessment of credit risk model adequacy is usually based on the use of statistical metrics of discriminatory power between risk classes, often referred as model validation, as well as on the forecasting of the empirically observed default frequency, often referred as model calibration. The most popular measures of discriminatory power constitute the Cumulative Accuracy Profile (CAP) and the Receiver Operating Characteristic (ROC), which can be summarized by the statistical metrics of Accuracy Ratio (AR) and Area under ROC (AUROC), respectively. Related statistics, but with limited applications, constitute the Kolmogorov–Smirnov test, the Mahalanobis distance as well as the Gini coefficient.

In this chapter, we present the advantages and disadvantages of various validation metrics, the degree of their information overlap, as well as the conditions under which such measures could lead to valid model assessments. A critical aspect in this direction is the recognition of randomness, which results in random estimators of validation measures. In

* The views expressed in this paper are those of the authors and should in no part be attributed to the Bank of Greece.
† Manchester Business School and Bank of Greece, Manchester, UK
‡ Trinity College and Faculty of Economics, University of Cambridge, Cambridge, UK

this respect, the assessment of two competing models should not be based on the absolute difference of the estimated validation measures, but rather on the statistical significance of this difference.

2. Measures of discriminatory power

The discrimination of debtors by using quantitative scoring models essentially intends to reflect the difference between true default and true non-default events. Thus, the model might lead to a correct or a wrong prediction of each state. In particular, the four possibilities correspond to 'correct alarm', 'correct non-alarm', 'false alarm' and 'false non-alarm', where the latter two correspond to Type-II and Type-I error, respectively. In percentage form, the above four possibilities are presented in Table 3.1.

2.1. The contingency table

In its static form, Table 3.1 is called contingency table. The default forecast row of the table contains the number of creditors who were correctly classified as defaulters as a percentage of the true number of defaults, as well as the number of creditors who were wrongly classified as defaulters – whilst they have survived – as a percentage of the true number of non-defaults (Type-II error). The non-default forecast row of the table contains the number of creditors who were wrongly classified as non-defaulters – whilst they have defaulted – as a percentage of the true number of defaults (Type-I error), as well as the number of creditors who were correctly classified as non-defaulters as a percentage of the true number of non-defaults. Clearly, each column sums up to unity; thus, knowledge of one element immediately implied knowledge of the other.

The contingency table reveals the complete static picture of model performance. However, the comparability of contingency tables from competing models would depend on the definition of default. As long as the model assessment concerns only the two basic classes, default and non-default, the contingency tables would be immediately comparable. In the case of more than two or continuous classes, e.g. from 1 to 100, the definition of default would depend on the subjective cut-off point of the classification scale into two categories. For different cut-off points, any model would exhibit different performance; thus, contingency tables could be used as a mean of assessment of competing models only for common cut-off points.

Table 3.1 Contingency table

	Default	Non-default
Default forecast	$\dfrac{Correct\ alarms}{Defaults}\%$	$\dfrac{False\ alarms}{Non\text{-}defaults}\%$
Non-default forecast	$\dfrac{False\ non\text{-}alarms}{Defaults}\%$	$\dfrac{Correct\ non\text{-}alarms}{Non\text{-}alarms}\%$
Total	100	100

2.2. The ROC curve and the AUROC statistic

A way to overcome the above difficulties of contingency tables would be to describe graphically the model performance for all possible cut-off points. Thus, the ROC curve is defined as the plot of the non-diagonal element combinations of a contingency table for all possible cut-off points. That is, the plot of correct alarm rate (CAR)

$$CAR = \frac{Correct\ alarms}{Defaults}\%$$

on the vertical axis, versus the false alarm rate (FAR)

$$FAR = \frac{False\ alarms}{Non\text{-}defaults}\%$$

on the horizontal axis, for all possible cut-off points. In Figure 3.1, we plot three possible ROC curves:

A perfect model would correctly predict the full number of defaults and would be represented by the horizontal line at the 100% correct alarm level. On the other side, a model with zero predictive power would be represented by the 45° straight line. Finally, any other case of some predictive power would be represented by a concave curve positioned between the two extreme cases. The complements of the vertical and the horizontal axes would then represent the diagonal elements of the contingency table. Thus, the complement of the vertical axis would give the false non-alarm rate, whilst the complement of the horizontal axis would give the correct non-alarm rate.

In the case that the ROC curve of a particular model lies uniformly above the ROC curve of a competing model, it is clear that the former exhibits superior discriminatory power for all possible cut-off points. In the case that the two curves intersect, the point of intersection shows that at that point the relative discriminatory of the models is reversed (see Figure 3.2).

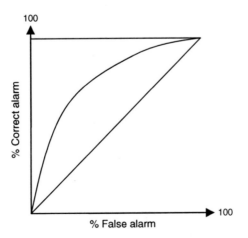

Figure 3.1 The receiver operating characteristic (ROC) curve

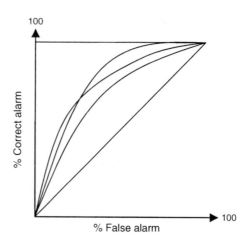

Figure 3.2 Alternative receiver operating characteristic (ROC) curves

The area under the ROC curve (AUROC) summarizes the information exhibited by the curve in a single statistic defined as

$$AUROC = \int_0^1 CAR\,(FAR)\,dFAR$$

The AUROC statistic is often used instead of ROC curve for the assessment of model performance. Clearly, the statistic takes values in the [0.5,1] interval where the two bounds correspond to models with zero and full discriminatory power, respectively. It must be emphasized that in case of two competing models, say 1 and 2, with intersecting ROC curves, even if $AUROC_1 > AUROC_2$, there will be a range of cut-off points in which model 2 might exhibit superior discriminatory performance. Therefore, in certain applications and selected cut-off points, the analyst might tend to approve models with smaller AUROC as compared with competing models.

2.3. The CAP curve and the AR statistic

The CAP curve is composed of combinations of the CAR (vertical axis) and the total number of defaults and non-defaults (horizontal axis) for all possible cut-off points. Graphically, we obtain the result shown in Figure 3.3.

A model of zero discriminatory power would be represented by the straight line of 45°, whilst a perfect model of full discriminatory power would increase linearly its CAR to 100%. Any other model of non-zero discriminatory power would be represented by a concave curve positioned between the two extreme cases.

A summary statistic for the information exhibited by a CAP curve is called AR. It is defined as the ratio of the area under the model's CAP curve over the area under the CAP curve of the perfect model and takes values in the [0.5,1] interval where the two bounds correspond to models with zero and full discriminatory power, respectively.

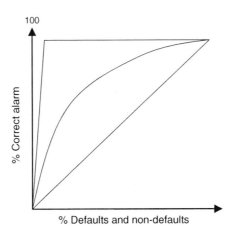

Figure 3.3 The cumulative accuracy profile (CAP) curve

There are arguments on the relationship between AUROC and AR and a study by Engelmann *et al.* (2003) shows that these statistics are essentially equivalent. In particular, it can be shown that

$$AR = 2\,(AUROC - 0.5)$$

Clearly, the AR is simply a linear transformation of the AUROC; thus, both statistics contain he same information and knowledge of any one of them would immediately imply knowledge of the other.

3. Uncertainty in credit risk model validation

Credit scoring models intend to discriminate between default and non-default events using a sample drawn from a non-observable population. The sample is composed of default and non-default events and the full set of scores estimated by a credit institution is an effort to infer the population properties from those observed in the sample. Given a number of model imperfections, the credit-scoring results are often subject to substantial variation because of the quality of the data sample and its composition between default and non-default events. Experimental empirical applications on extended databases, see Stein (2002), have shown that the existence of a small number of defaults in a data sample constitutes the main factor affecting the variation of model discriminatory performance. In real-world applications, most credit institutions have limited access to adequate and valid databases, and in this sense, they are subject to extensive risk of under- or over-assessment of credit risk model adequacy. On this basis, credit institutions should emphasize on the analysis and comprehension of scoring performance variation so that model validation is interpreted adequately. The literature has proposed a number of experts but also simple and applicable methods in this direction, which are often borrowed from other disciplines, see Pesaran and Skouras (2001).

3.1. Confidence intervals for AUROC

Let us consider every credit score as one possible realization of a random variable. In the case that the scoring concerns an element from our sample of defaults, this is denoted as R_D, otherwise R_{ND}. Thus, the probability to characterise correctly a default event is $Pr(R_D < R_{ND})$. Because for every cut-off point, C, the CAR and the FAR are defined as

$$CAR = Pr\,(R_D < C)$$

$$FAR = Pr\,(R_{ND} < C)$$

then,

$$Pr\,(R_D < R_{ND}) = AUROC$$

Mann and Whitney (1947) have established the U-statistic for the static test of zero discriminatory power hypothesis

$$Pr\,(R_D < R_{ND}) = 0.5$$

of a given model. The U-statistic is essentially a statistically unbiased estimator of AUROC. On the basis of Mann and Whitney, Engelmann $et\ al.$ (2003) have shown that the U-statistic is asymptotically normally distributed and propose the calculation of a confidence interval of the form

$$\left[\hat{U} - \hat{\sigma}_{\hat{U}}\Phi^{-1}\left(\frac{1+\alpha}{2}\right),\ \hat{U} + \hat{\sigma}_{\hat{U}}\Phi^{-1}\left(\frac{1+\alpha}{2}\right) \right]$$

The true value of AROC should be within the confidence interval with probability α, whilst $\hat{U}$, $\hat{\sigma}_{\hat{U}}$ and Φ denote the estimated U-statistic, its estimated standard deviation and the standard normal distribution, respectively. Empirical analysis suggests that the above confidence interval is a satisfactory proxy as long as the number of defaults in the sample is at least 50, otherwise one may use alternative computational methods such as those described in Bamber (1975).

Clearly, the difference in the estimated value of the AUROC statistic of two competing scoring models can be due to chance. DeLong $et\ al.$ (1988) as well as Engelmann $et\ al.$ (2003) show that it is possible to test for the statistical significance of the difference between AUROC estimates from competing models. In particular, we can show that the standardized difference between the U-statistics of two models

$$\frac{\left(\hat{U}_1 - \hat{U}_2\right)^2}{\sigma_{\hat{U}_1}^2 + \sigma_{\hat{U}_2}^2 - 2\sigma_{\hat{U}_1\hat{U}_2}}$$

follows a $X^2\,(1)$ distribution, where $\sigma_{\hat{U}_1}^2$, $\sigma_{\hat{U}_2}^2$, $\sigma_{\hat{U}_1\hat{U}_2}^2$ denote the variances of the U-statistics from models 1 and 2 as well as their covariance. Therefore, we can immediately test for the statistical significance of the difference between AUROC estimates from competing models.

3.2.   The Kupiers Score and the Granger–Pesaran Test

The Kupiers Score (KS) was originally used in the evaluation of weather forecasts, see Murphy and Dann (1985), as well as in the construction of Henrikson and Merton (1981) test statistic for market-timing in finance, which was subsequently generalized by Pesaran and Timmermann (1992). It is defined as the distance between the CAR and the FAR:

$$KS = CAR - FAR$$

Granger and Pesaran (2000), show that the Pesaran–Timmermann statistic can be expressed as

$$GP = \frac{\sqrt{N} \times KS}{\left(\frac{p_f(1-p_f)}{p_a(1-p_a)}\right)^{\frac{1}{2}}} \overset{a}{\sim} N(0,1)$$

where p_a denotes the number of realized defaults over the full number of data (default and non-default) and p_f denotes the number of correctly and incorrectly predicted defaults over the full number of data (default and non-default). The GP statistic can be shown to follow a standard normal distribution, thus making testing procedures readily available.

4.   Confidence interval for ROC

Basel II recommendations place the validation of credit risk models at the heart of credit risk management processes. Financial institutions develop credit-rating models that are often judged on the basis of statistical metrics of discriminatory power as well as default forecasting ability. The ROC is the most popular metric used by creditors to assess credit-scoring accuracy and as part of their Basel II model validation. Related tests but with relative limited applications are the Kolmogorov–Smirnov test, the Mahalanobis distance as well as the Gini coefficient. A recent review paper has been published by Tasche (2005). Christodoulakis and Satchell (2006) developed a model to provide a mathematical procedure to assess the accuracy of ROC curve estimates for credit defaults, possibly in the presence of macroeconomic shocks, supplementing the non-parametric method recommended by Engelmann et al. (2003) based on the Mann and Whitney (1947) test. In this section, we shall focus on the properties of the method offered by Christodoulakis and Satchell (2006).

4.1.   Analysis under normality

Let us denote sick and healthy credit scores by y and x, respectively. Both are assumed to be absolutely continuous random variables, and their distribution functions are defined as $F_y(c) = \Pr(y \leq c)$ and $F_x(c) = \Pr(x \leq c)$, respectively, and $F_x^{-1}(.)$ denotes the inverse distribution function that is uniquely defined. A perfect rating model should completely separate the two distributions whilst for an imperfect (and real) model perfect discrimination is not possible and the distributions should exhibit some overlap. The latter situation is presented in Figure 3.4 using normal density functions.

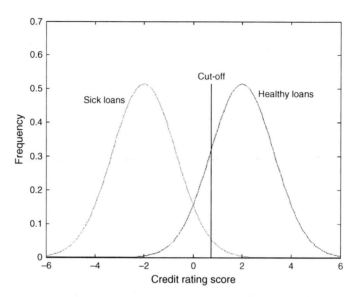

Figure 3.4 Rating score distributions for sick and healthy loans (*Source*: Christodoulakis and Satchell, 2006)

The decision maker discriminates the debtors belonging to the two classes by introducing a cut-off point as depicted in Figure 3.4, which would classify all the debtors below that point as potential defaulters and those above as potential survivors. We then have four possible decision results as clearly described by Sobehart and Keenan (2001): (1) debtors classified below cut-off that eventually defaulted (correct alarms), (2) debtors classified below cut-off that eventually survived (false alarms), (3) debtors classified above cut-off that eventually survived (correct survivors) and (4) debtors classified above cut-off that eventually defaulted (missed alarms).

We can then construct the ROC by calculating for every possible cut-off point in the range of rating scores, the ratio of correct alarms to total number of defaults (CAR) and false alarms to total number of non-defaults (FAR). Then, ROC is defined as the plot of pairs of CAR versus FAR. Clearly, both quantities take values between zero and one, and in Figure 3.1, CAR can be represented by the integral of the sick loan density up to the cut-off point whilst FAR can be represented by the integral of the healthy loan density up to the cut-off point. This probabilistic interpretation leads us to state the following proposition, which is taken from Christodoulakis and Satchell (2006).

Proposition 1
If the credit-rating scores for defaulters y and non-defaulters x are represented by mutually independent normally distributed random variables $y \sim N\left(\mu_y, \sigma_y^2\right)$ and $x \sim N\left(\mu_x, \sigma_x^2\right)$, respectively, then the ROC satisfies the following relationship:

$$CAR = \Phi\left(\Phi^{-1}\left(FAR\right)\right) = \Phi\left(\frac{\left(\mu_x - \mu_y\right) + \Phi^{-1}\left(FAR\right)\sigma_x}{\sigma_y}\right)$$

where $\Phi(\ )$ denotes the cumulative standard normal density.

Proof See Christodoulakis and Satchell (2006).

We can estimate the unknown parameters $\mu_x, \mu_y, \sigma_x, \sigma_y$, as the usual sample moments, given two samples $(x_1, \ldots, x_{N_1})$ and $(y_1, \ldots, y_{N_2})$.

4.2. Examples under normality

Let us consider some numerical examples presented in the original paper to illustrate the relevance of estimated parameters in our analysis. We assume a 'true' data-generating process in which the means of sick and healthy loan credit-rating scores are −7 and 2, respectively, and their standard deviation is 5. Also, assume that sick (healthy) loan mean has been underestimated (overestimated) taking value −8 (3). This would result in a false belief that the rating model exhibits superior performance over the entire range of FARs. The reverse results would become obvious in the case that sick (healthy) loan mean had been overestimated (underestimated) taking value −6 (1). We plot all three cases in Figure 3.5.

Then, let us assume that sick (healthy) loan standard deviation has been underestimated (overestimated) taking value 4 (7). This would result in a false belief that the rating model exhibits superior (inferior) performance for low (high) FARs. The reverse results would become obvious in the case that sick (healthy) loan standard deviation had been overestimated (underestimated) taking value 7 (4). We plot all three cases in Figure 3.6.

4.3. The construction of ROC confidence intervals

In this subsection, we shall turn our attention to the construction of ROC confidence intervals. Let us denote our estimated y by $CAR\left(\hat{\theta}, FAR\right)$ versus the true $CAR\left(\theta, FAR\right)$, where $\theta = \left(\mu_x, \mu_y, \sigma_x, \sigma_y\right)$. Then, from Christodoulakis and Satchell (2006), we state the following proposition for the confidence interval of CAR.

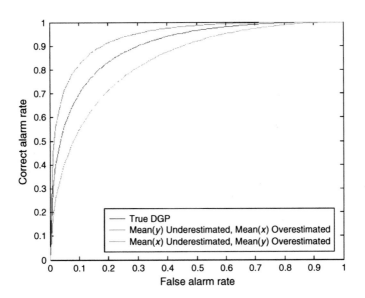

Figure 3.5 ROC sensitivity to mean mis-estimation (*Source*: Christodoulakis and Satchell, 2006)

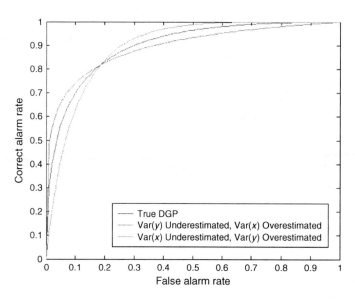

Figure 3.6 ROC sensitivity to variance mis-estimation (*Source*: Christodoulakis and Satchell, 2006)

Proposition 2
If the credit-rating scores for defaulters y and non-defaulters x are represented by mutually independent normally distributed random variables $y \sim N\left(\mu_y, \sigma_y^2\right)$ and $x \sim N\left(\mu_x, \sigma_x^2\right)$, respectively, and Proposition 1 holds, then the confidence interval for CAR is given by

$$CAR\left(\hat{\theta}, FAR\right) - \Phi^{-1}\left(\frac{1+alpha}{2}\right)d < CAR\left(\theta, FAR\right)$$

$$< CAR\left(\hat{\theta}, FAR\right) + \Phi^{-1}\left(\frac{1+alpha}{2}\right)d \quad (3.6)$$

where $\Phi^{-1}\left(\frac{1+alpha}{2}\right)$ is the upper *alpha* per cent point of the standard normal and

$$
\begin{aligned}
d^2 &= \frac{\varphi(a)^2}{\sigma_y^2}\begin{pmatrix}1\\-1\\b\\-a\end{pmatrix}'\begin{pmatrix}\sigma_x^2 & & & \\ & \sigma_y^2 & & \\ & & \frac{\sigma_x^2}{2} & \\ & & & \frac{\sigma_y^2}{2}\end{pmatrix}\begin{pmatrix}1\\-1\\b\\-a\end{pmatrix} \\
&= \frac{\phi^2(a)}{2\sigma_y^2}\left(2\sigma_x^2 + 2\sigma_y^2 + b^2\sigma_x^2 + a^2\sigma_y^2\right) \\
&= \frac{\exp\left(-a^2\right)}{\pi\sigma_y^2}\left(2\sigma_x^2 + 2\sigma_y^2 + b^2\sigma_x^2 + a^2\sigma_y^2\right)
\end{aligned}
\quad (5)
$$

where

$$h = \Phi^{-1}(x)$$
$$a = \frac{(\mu_x - \mu_y) + h\sigma_x}{\sigma_y}$$

Proof See Christodoulakis and Satchell (2006).

4.4. Analysis under non-normality

In this subsection, we shall study the effects of non-normality by assuming that our data are generated by independent skew-normal distributions, originally introduced by O'Hagan and Leonard (1978) as priors in Bayesian estimation and developed by Azzalini (1985, 1986) and further generalized by Azzalini and Dalla Valle (1996) and Arnold and Lin (2004) among others. Let us denote

$$y = \mu_y + \sigma_y v_y$$

where $v_y \sim SN(\lambda_y)$ and λ_y is a real constant then the density function of the skew normal distribution for v_y is given by

$$pdf(v_y) = 2\varphi(v_y) \Phi(\lambda_y v_y)$$

where

$$\varphi(v_y) = \frac{1}{\sqrt{2\pi}} \exp\left(-\frac{v_y^2}{2}\right)$$

is the standard normal density and $\Phi()$ is the cumulative standard normal. In this context, the parameters μ, σ and λ refer to location, scale and skewness parameters, respectively, and do not correspond to moments. The skew normal accommodates a variety of skewness patterns as λ varies, whilst it converges to the Normal as $\lambda \to 0$. Similarly, for non-default data we assume

$$x = \mu_x + \sigma_x v_x$$

where $v_x \sim SN(\lambda_x)$. We can show that

$$E(x) = \mu + \sigma\lambda \sqrt{\frac{2\pi}{1 + \lambda^2}}$$
$$\mathrm{Var}(x) = \sigma^2 \left(1 - \frac{2\lambda^2}{\pi(1 + \lambda^2)}\right)$$

whilst both skewness and kurtosis also depend on λ. Thus, in the presence of skew-normal data-Generating processes, the decision problem of discriminating between default and non-default scoring distributions, as depicted in Figure 3.1, would have a large number of analogues depending on the relative values of skewness parameters λ_x and λ_y on top of location and scale parameters.

4.5. Examples under non-normality

Adopting the examples of Christodoulakis and Satchell (2006), for $\lambda_y = 3$ and $\lambda_x = -3$ the likelihood of making discrimination errors is shown to decrease in Figure 3.7, but when $\lambda_y = -3$ and $\lambda_x = 3$, we observe clearly that the distributions develop extensive overlap which in turn enhances the likelihood of making both types of discrimination errors (see Figures 3.7 and 3.8).

Our results under skew normal are now summarized in Proposition 3.

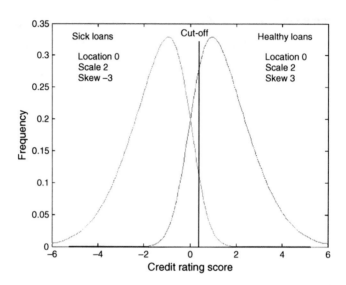

Figure 3.7 Rating score distributions for sick and healthy loans (*Source*: Christodoulakis and Satchell, 2006)

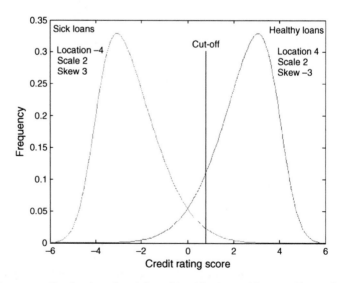

Figure 3.8 Rating score distributions for sick and healthy loans (*Source*: Christodoulakis and Satchell, 2006)

Proposition 3

If the credit-rating scores for defaulters y and non-defaulters x are represented by mutually independent skew-normally distributed random variables $y \sim SN\left(\mu_y, \sigma_y^2, \lambda_y\right)$ and $x \sim SN\left(\mu_x, \sigma_x^2, \lambda_x\right)$ respectively, then the ROC satisfies the following relationship.

$$CAR = CSN\left(\frac{\left(\mu_x - \mu_y\right) + CSN^{-1}\left(FAR; \lambda_x\right)\sigma_x}{\sigma_y}; \lambda_y\right)$$

$$= \Phi\left(\frac{\left(\mu_x - \mu_y\right) + CSN^{-1}\left(FAR; \lambda_x\right)\sigma_x}{\sigma_y}\right) - 2T\left(\frac{\left(\mu_x - \mu_y\right) + CSN^{-1}\left(FAR; \lambda_x\right)\sigma_x}{\sigma_y}; \lambda_y\right)$$

where CSN( ) and Φ( ) denote the cumulative skew-normal and standard normal densities, respectively, and T( ) denotes the Owen (1956) function.

Proof   See Christodoulakis and Satchell (2006).

Note that given values of g and k, the Owen (1956) function $T(w, k)$ calculates the quantity

$$T\left(w, k\right) = \frac{1}{2\pi}\int_0^k d\frac{\exp\left(-\frac{w^2}{2}\left(1 + x^2\right)\right)}{1 + x^2}dx$$

The ROC curve described in Proposition 3 has a more general form as compared with that of Proposition 1 in that it is affected not only by location and scale parameters but also by the shape parameter. This allows for further flexibility and accuracy in generating ROC curves as we can show that the four moments of the skew-normal distribution are all affected by the presence of skewness. Let us consider a 'true' data-generating process in which the means of sick and healthy loan credit-rating scores are -7 and 2, respectively, and their standard deviation is 5. In addition, the true sick and healthy loan shape parameters are 1 and 0.3, respectively. Let us assume that sick loan shape parameter has been mis-estimated taking possible values 0, 1.5 and 2.5, respectively. Plotting these alternative ROC curves in Figure 3.9, we observe clearly that sick skewness underestimation (overestimation) results in a false belief of rating model superior (inferior) performance over the entire range of FARs. Ultimate under- (over) estimation of skewness parameter would lead the analyst to the false conclusion that the model approaches perfectly efficient (inefficient) performance.

In the case that both sick and healthy loan parameters are mis-estimated or under different true data-generating processes, these comparative static effects would be effectively altered. That is, using all the parameter values as described above, but for healthy loan skewness 1.3 or -1.3, our resulting ROC would be depicted as in Figures 3.10 and 3.11, respectively.

In Figures 3.10 and 3.11, we see that our false impression on the performance of a model is subject to shape parameter trade-offs between sick and healthy loan distributions.

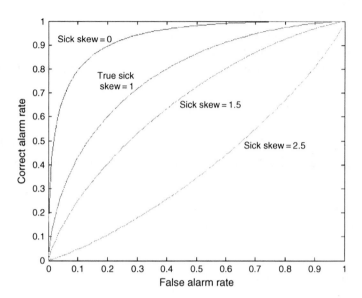

Figure 3.9 ROC sensitivity to sick loan skew mis-estimation (*Source*: Christodoulakis and Satchell, 2006)

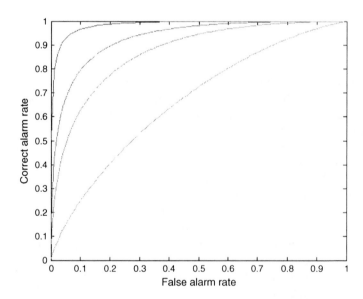

Figure 3.10 ROC sensitivity to sick loan skew mis-estimation (*Source*: Christodoulakis and Satchell, 2006)

Of course, the picture would further complicate in case that location and scale parameters change as well. The following corollary describes the relationship between Proposition 1 and Proposition 3.

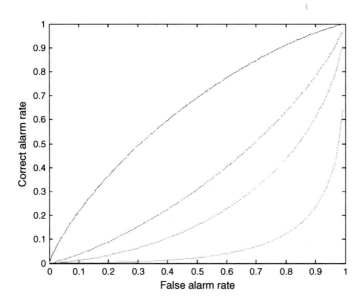

Figure 3.11 ROC sensitivity to sick loan skew mis-estimation (*Source*: Christodoulakis and Satchell, 2006)

Corollary
The ROC of Proposition 1 is a special case of Proposition 3.

Proof See Christodoulakis and Satchell (2006).

5. Bootstrapping

A common method to empirically extract statistics with unknown properties in the presence of small samples is called Bootstrapping, which is widely used in the area of econometrics and statistics. In brief, suppose the analyst has available a sample of scoring results from which he/she randomly draws a sub-sample. Using the sub-sample, he/she then calculates the chosen statistic, e.g. AUROC or AR, and then records the resulting value. In the following, the analyst replaces the sub-sample in the original sample of data and applies the same process for a large number of iterations, and in every iteration, the value of the statistic under investigation is recorded. Thus, an empirical distribution of the statistic will become available, and therefore, an assessment of its uncertainty will be readily available. This process is possible to provide with satisfactory results under generally acceptable conditions, see Efron and Tibshirani (1993).

6. Optimal rating combinations

Building confidence intervals and robust standard errors for model validation statistics is a natural way to assess their variability. A possible reduction in variability could be attained through the use of successful methods developed in the area of econometrics as well as the financial theory of portfolio selection. For a number of debtors, it is possible to observe scoring results both from external rating agencies as well as from internal rating systems. These alternative scoring results often differ, a phenomenon that is often called 'split ratings'. To some extent, split ratings reflect some common information that is publicly available, plus some information that is exclusive or subjective to every different scoring system or analyst. Thus, it might be possible to create a ratings combination for each debtor, which is a portfolio of ratings, which could exhibit smaller variation as compared with each individual rating. These combinations are called 'synthetic' or 'composite' ratings. From a technical point of view, this is constructed in a way similar to the construction of optimal portfolios of assets, as long as the measure of risk for the user is the variance. For an application in this context, see Tabakis and Vinci (2002). However, we should emphasize that the construction of minimum variance synthetic ratings reflect symmetric preferences with respect to the cost from over-rating versus under-rating. From the point of view of a prudent bank supervisor and credit institutions, the cost of imperfect scoring should be asymmetric, where over-scoring should be more costly as compared with equal under-scoring. In this case, minimum variance composite ratings should yield inferior results. The standardized form of Basel II proposes the adoption of the median in case of split ratings, thus encouraging the formation of composite ratings.

7. Concluding remarks

The validation of credit risk models constitutes an indispensable element in the process of quantitative risk management. The comparative evaluation of competing models is largely based on the use of statistical measures such as the CAP and ROC curves and the respective statistics AR and AUROC, which constitute popular choices. In this chapter, we have examined the conditions under which the use of such methods would yield valid scoring results. We show that such metrics are subject to variation, and thus, the analyst should be interested not in the absolute difference between statistics, but in the statistical significance of this difference. We have analysed a number of alternative approaches dealing with this problem.

References

Arnold, B.C. and Lin, G.D. (2004). Characterizations of the skew-normal and generalized chi distributions. *Sankhyā: The Indian Journal of Statistics*, 66, 593–606.

Azzalini, A. (1985). A class of distributions which includes the normal ones. *Scandinavian Journal of Statistics*, 12, 171–8.

Azzalini, A. (1986). Further results on a class of distributions which includes the normal ones. *Statistica*, 46, 199–208.

Azzalini, A. and Dalla Valle, A. (1996). The multivariate skew-normal distributions. *Biometrika*, 83, 715–26.

Bamber, D. (1975). The area above the ordinal dominance graph and the area below the receiver operating graph. *Journal of Mathematical Psychology*, 12, 387–415.

Christodoulakis, G.A. and Satchell, S.E. (2006). Assessing the accuracy of credit R.O.C. estimates in the presence of macroeconomic shocks, Working Paper, Manchester Business School.

DeLong, E., DeLong, D., and Clarke-Pearson, D. (1988). Comparing the areas under two or more correlated receiver operating characteristic curves: a non parametric approach. *Biometrics*, 44, 837–45.

Efron, B. and Tibshirani, R. (1993). An introduction to the bootstrap. Chapman and Hall/CRC Press, Monographs on Statistics and Applied Probability, No 57.

Engelmann, B., Hayden, E. and Tasche, D. (2003). Testing rating accuracy. *Risk*, 16, 82–6.

Granger, C.W.J. and Pesaran, M.H. (2000). Economic and statistical measures of forecast accuracy. *Journal of Forecasting*, 19, 537–60.

Henrikson, R.D. and Merton, R.C. (1981). On market-timing and investment performance II: Statistical procedures for evaluating forecasting skills. *Journal of Business*, 54, 513–33.

Mann, H. and Whitney, D. (1947). On a test of whether one or two random variables is stochastically larger than the other. *Annals of Mathematical Statistics*, 18, 50–60.

Murphy, A.H. and Dann, H. (1985). Forecast evaluation. In *Probability, Statistics and Decision Making in the Atmospheric Sciences* (A.H. Murphy and R.W. Katz, eds) Westview, Boulder, pp. 379–437.

O'Hagan, A. and Leonard, T. (1978). Bayes estimation subject to uncertainty about parameter constraints. *Biometrika*, 63, 201–3.

Owen, D.B. (1956). Tables for computing bivariate normal probabilities. *Annals of Mathematical Statistics*, 27, 1075–90.

Pesaran, M.H. and Skouras, S. (2001). Decision-based methods for forecast evaluation. In *A Companion to Economic Forecasting* (M.P. Clements and D.F. Hendry, eds) Blackwell, Oxford, pp. 241–67.

Pesaran, M.H. and Timmermann, A. (1992). A simple non-parametric test of predictive performance. *Journal of Business and Economic Statistics*, 10, 461–5.

Sobehart, J. and Keenan, S. (2001). Measuring default accurately. *Risk*, 11, S31–3.

Stein, R. (2002). Benchmarking default prediction models: pitfalls and remedies in model validation. Technical Report No 020305, Moody's KMV, New York.

Tabakis, E. and Vinci, A. (2002). *Analysing and Combining Multiple Credit Assessments of Financial Institutions*. Working Paper No 123, European Central Bank.

Tasche, D. (2005), Rating and probability of Default Validation. In *Studies in the Validation of Internal Rating Systems*. Working Paper No 14, Basel Committee on Banking Supervision, BIS.

4 A moments-based procedure for evaluating risk forecasting models

*Kevin Dowd**

Abstract

This chapter examines the important problem of evaluating a risk forecasting model [e.g. a values-at-risk (VaR) model]. Its point of departure is the likelihood ratio (LR) test applied to data that have gone through Probability Integral Transform and Berkowitz transformations to become standard normal under the null hypothesis of model adequacy. However, the LR test is poor at detecting model inadequacy that manifests itself in the transformed data being skewed or fat-tailed. To remedy this problem, the chapter proposes a new procedure that combines tests of the predictions of the first four moments of the transformed data into a single omnibus test. Simulation results suggest that this omnibus test has considerable power and is much more robust than the LR test in the face of model misspecification. It is also easy to implement and does not require any sophistication on the part of the model evaluator. The chapter also includes a table giving the test bounds for various sample sizes, which enables a user to implement the test without having to simulate the bounds themselves.

1. Introduction

There has been a lot of interest in the last few years in risk forecasting (or probability-density forecasting) models. Such models are widely used by financial institutions to forecast their trading, investment and other financial risks, and their risk forecasts are often a critical factor in firms' risk management decisions. Given the scale of their reliance on them, it is therefore important for firms to ensure that their risk models are properly evaluated. Any such evaluation must involve some comparison of time series of both forecast density functions and subsequently realized outcomes, but the comparison is complicated by the fact that forecast densities typically vary over time. Forecast evaluation is also made more difficult by the limited sizes of the data sets typically available, and it is well known that standard interval forecast tests often have very low power with real world sample sizes.

This chapter offers a new approach to the problem of risk model evaluation. It builds on earlier work in particular by Crnkovic and Drachman (1995), Diebold *et al.* (1998) and Berkowitz (2001). The first two of these studies suggested how the problem of evaluating a density function that changes over time can be 'solved' by mapping realized outcomes to their percentiles in terms of the forecasted density functions. This Probability Integral

* Nottingham University Business School, Jubilee Campus, Nottingham, UK

Transform (PIT) transformation makes it possible to evaluate density forecasting models by testing whether the distribution of mapped outcomes is consistent with the uniform distribution predicted under the null that the risk model is adequate. Berkowitz then suggested that these mapped observations should be run through a second transformation to make them into a series z that is distributed as standard normal under the null. A major attraction of this suggestion is that it enables the model evaluator to make use of the wide variety of normality tests available. Testing for model adequacy then reduces to testing whether z is distributed as standard normal. Berkowitz focused on the likelihood ratio (LR) test for this purpose, and his results suggested that it was superior to standard interval forecast approaches such as those suggested by Kupiec (1995) or Christoffersen (1998). His results also suggested that the LR test had good power with the sample sizes often available for model validation purposes.

This chapter presents a more pessimistic view of the power of the LR test, backed up by simulation results showing the power of the LR test against a wide range of specified departures from standard normality. The problem with the LR test is that it only examines a subset of the 'full' null hypothesis – more specifically, it tests whether z has a mean of 0 and a standard deviation of 1. However, it is quite possible for z to satisfy these predictions and yet be non-normally distributed, and in such cases, the LR test would typically overlook model inadequacy. Thus, the LR test focuses on whether the first two moments of z are compatible with standard normality but has little power in the face of departures from standard normality that manifest themselves in the higher moments of z.

More constructively, this chapter also suggests a new testing approach that is reliable in the sense of having good power against a wide range of departures from standard normality. The proposed test is a composite test that takes equal account of all first four moment predictions and tests whether each sample moment is consistent with what we would expect it to be under standard normality. The test can therefore be regarded as an omnibus test of model adequacy. The test also has considerable power against a wide range of possible departures from standard normality, including those that manifest themselves in the third and fourth moments of z. The new test is therefore more reliable than the LR test for model validation purposes.

Furthermore, the test is intuitive and does not require a high level of statistical or econometric expertise on the part of the model evaluator who implements it. It is also easy to carry out and can be implemented quickly on a simple spreadsheet. These features make it ideal for model evaluators who are not specialist statisticians or econometricians, and this would include many risk managers, regulatory supervisors or auditors faced with the task of signing off risk models as part of their routine risk management, supervisory or audit functions.

This chapter is laid out as follows. Section 2 goes through some preliminary mapping analysis and shows how the original P/L data can be mapped to the z series that is standard normal under the null hypothesis that the model is adequate. Section 3 examines the use of an LR test to evaluate the standard normality prediction and presents some results showing the power of the LR test against a variety of departures from standard normality. Section 4 then explains the moments-based test, discusses its implementation, and presents results showing its power against the same non-standard normality distributions considered in Section 3. Section 5 presents a worked example illustrating how the new test can detect the inadequacy of a risk model that under-estimates the heaviness of the tails of the forecasted distribution, which is a common problem with many values-at-risk (VaR) models. Section 6 concludes.

2. Preliminary analysis

Suppose we have a risk model that forecasts the density function for a random variable over a specified forecast period. Typically, the model might forecast a financial institution's VaR over the next trading day, and the random variable might be the daily profit/loss. The task is to evaluate the model's density forecasting performance over a sample period. The evaluator might be a risk manager employed by the institution, the institution's regulatory supervisor or a risk auditor.

We also assume that the evaluator knows the density forecasts. This assumption makes sense in the VaR context because any of these possible evaluators would have access to this information in practice. This assumption also makes sense in some other contexts – for instance, it would make sense where the density forecasts are published (e.g. as with the inflation density forecasts of the Survey of Professional Forecasters or the UK inflation density forecasts published by the Bank of England since 1997) or based on model-free forecasts [e.g. as with model-free approaches to estimating risk neutral densities of the underlying asset in an option, as suggested by Aït-Sahalia and Lo (1998)].[1]

To evaluate the model's density forecasts, we carry out a PIT transformation and map each realized value to its percentile point on its forecasted density function (e.g. if the realized value over some period is equal to the 90-th percentile of its forecasted density function, then this value maps to 0.90, etc.). Now let p be the realized values mapped to their percentiles on the forecasted densities.

Under the null hypothesis that the model is adequate, the mapped series p should be uniformly distributed over the interval $(0,1)$, so we can evaluate the model directly by testing the prediction that p is uniformly distributed (e.g. using a Kolmogorov test). However, following Berkowitz (2001), testing is easier if we transform p into a series that is standard normal under the null. We can do so by applying a standard normal inverse function to the p series, viz.:

$$z = \Phi^{-1}(p) \tag{4.1}$$

This second, Berkowitz, transformation is helpful because we have a wider and more powerful range of tests available to us if we deal with standard normal random variables than if we deal with standard uniform ones. We can therefore evaluate the model's adequacy by testing the prediction that the PIT–Berkowitz transformed data should be standard normal.[2]

3. The likelihood ratio test

One way to do so is by means of a LR test, as suggested by Berkowitz (2001). However, such a test only addresses part of the 'full' null hypothesis – specifically, it tests the predictions that the mean μ and standard deviation σ of z are 0 and 1, respectively, but does not test that the distribution of z is normal. To emphasize the distinction, note that the 'full' null, $H_0 : \{z \sim N(0,1)\}$, can be decomposed into two separable sub-hypotheses $H_0^a : \{\mu = 0, \sigma = 1\}$ and $H_0^b : \{z \sim N(\mu,\sigma^2)\}$. We might expect the LR test to be good at detecting departures from the null that arise from departures from H_0^a, but we would not necessarily expect it to have much power against departures from the null that arise from departures from H_0^b.

To investigate further, Table 4.1 shows power results for the LR test against a considerable variety of illustrative departures from the full standard normal null hypothesis. The first three lines give results for illustrative departures from H_0^a involving either or both of $\mu \neq 0$ and $\sigma \neq 1$. In each of these cases, z is normal, so there is a departure from H_0^a but not from H_0^b. The next eight lines show results for illustrative departures from H_0^b: three cases of skewness but no leptokurtosis, three cases of leptokurtosis but no skewness and two cases involving both skewness and leptokurtosis. In each of these cases, there is a departure from H_0^b but no departure from H_0^a, so $\mu = 0$ and $\sigma = 1$. Each line gives a description of the particular distribution, the values of its first four moments and power results for sample sizes of 50, 100, 250, 500 and 1000. These latter results are obtained using Monte Carlo simulation with 10 000 trials.[3]

The results in Table 4.1 show that our expectations about the LR test are confirmed. The LR test is powerful against departures from H_0^a, and (as we would also expect) its power rises with the sample size. For example, with a sample size $n = 1000$, the power of the test against each of the three specified departures from standard normality is 0.810, 0.973 and 0.998, respectively. (The significance level or type I error rate is set at 0.05.) However, the remaining results indicate that the LR test typically has little or no power against alternative hypotheses involving departures from normality. Indeed, in most cases, the power of the test remains close to the significance level regardless of the sample size, which indicates that the LR test has little or no discriminatory power against the particular departure concerned. Thus, the overall impression is that the LR test is good at detecting departures from H_0^a but is of little – and typically almost no – use in detecting departures from H_0^b.

4.   A moments test of model adequacy

We seek an alternative testing procedure that is more sensitive to departures from H_0^b whilst still retaining sensitivity to departures from H_0^a.[4] A good starting point is to consider the moment predictions implied by these hypotheses: in particular, H_0^a predicts that the first moment (i.e. that associated with the mean) should be 0 and the second moment (i.e. that associated with the variance) should be 1, and H_0^b predicts that the third moment (i.e. that associated with the skewness) should be 0 and the fourth moment (i.e. that associated with the kurtosis) should be 3.[5] Given that we want a robust testing framework that is sensitive to failures of any of these moment predictions, it makes sense to require that it give each of these moment predictions equal weight in the rejection process.[6]

Now imagine that we have a large number m of samples drawn from a standard normal, and each sample is of size n. Each sample can be regarded as a sample z series, and imagine that we put each of these samples through a sequence of four filters. The first filter applies a test of the first moment prediction and rejects a certain proportion of samples. Samples that pass the first test are put through a second filter that applies a test of the second-moment prediction and again rejects a certain proportion of samples. Those that survive this test are put through a test of the third-moment prediction and

Table 4.1 Power results for likelihood ratio test[1]

Departure from standard normality	Type of distribution	Mean	Std	Skew	Kurtosis	Power for sample size (n)				
						$n = 50$	$n = 100$	$n = 250$	$n = 500$	$n = 1000$
Alternative involves normal with non-standard parameter values										
Mean $\neq 0$	non-standard normal	0.1	1	0	3	0.097	0.132	0.271	0.499	0.810
Std $\neq 1$	non-standard normal	0	1.1	0	3	0.137	0.224	0.492	0.782	0.973
Mean $\neq 0$ and Std $\neq 1$	non-standard normal	0.1	1.1	0	3	0.182	0.315	0.648	0.925	0.998
Alternative involves $\mu = 0$, $\sigma = 1$ and non-normal distribution										
Skewed	Skew-t with $\nu = 100$ and $\lambda = 25^2$	0	1	0.11	≈ 3	0.060	0.055	0.057	0.062	0.074
Skewed	2PN with $\sigma_2 = 1.1^3$	0	1	0.16	≈ 3	0.056	0.055	0.052	0.052	0.053
Skewed	2PN with $\sigma_2 = 1.2^3$	0	1	0.32	≈ 3	0.056	0.056	0.054	0.056	0.058
Leptokurtic	Error distribution with $c = 1.25^4$	0	1	0	3.56	0.075	0.075	0.076	0.076	0.075
Leptokurtic	Generalized symmetric-t with $\nu = 15^5$	0	1	0	3.55	0.074	0.069	0.067	0.071	0.076
Leptokurtic	Generalized symmetric-t with $\nu = 5^5$	0	1	0	9	0.169	0.179	0.206	0.215	0.219
Skewed and leptokurtic	Skew-t with $\nu = 5$ and $\lambda = 3^2$	0	1	1	3.77	0.083	0.083	0.082	0.083	0.095
Skewed and leptokurtic	Skew-t with $\nu = 15$ and $\lambda = 5^2$	0	1	0.61	4.12	0.077	0.080	0.080	0.082	0.086

Notes:
1. Significance level (type I error rate) = 0.05, and each result is based on the outcome of 10 000 Monte Carlo simulation trials. Only results for positive departures from μ and σ nulls and positive skews are reported, as the corresponding results for negative departures and negative skews are similar.
2. The skew-t parameters are those of the (ν, λ) parameterization in Jones and Faddy (2003), with mean and standard deviation set to 0 and 1, respectively.
3. The two-piece normal (2PN) parameters are those of the $(\mu, \sigma_1, \sigma_2)$ parameterization of John (1982), with mean and standard deviation set to 0 and 1, respectively.
4. The error distribution parameters are as given in Evans *et al.* (2000), with mean and standard deviation set to 0 and 1, respectively.
5. The symmetric-t parameters are as given in Evans *et al.* (2000), with mean and standard deviation set to 0 and 1, respectively.

those that survive this are put through a test of the fourth-moment prediction. The poin
of this exercise is to obtain the bounds demarcating whether we accept or reject a sampl
at each stage. These are determined by the desired rejection rates, and it is these bound
that we use to implement our test.

To make the discussion concrete, suppose we wish to reject a proportion 0.01275 o
samples at each stage. Say $m = 50\,000$ and $n = 50$. We then estimate the sample mean
and find that the $1.275/2 = 0.6375$ and $100 - 1.275/2 = 99.3625$ percentile points o
the distribution of simulated sample means are -0.350 and 0.351. This tells us tha
98.73% of the simulated sample means lie between -0.350 and -0.351. (These value
are, of course, also the bounds of the central 98.73% confidence interval for the simulate
sample means.) We then eliminate those samples whose means lie outside this range. W
now estimate the variances of our remaining samples and find that 98.73% of these li
between 0.7554 and 1.256. We then eliminate those samples whose variances lie outsid
this range and estimate the skewness parameters of our remaining samples. Proceeding i
similar manner, we find that 98.73% of the estimated skewnesses lie between -0.850 anc
0.847. We then eliminate those samples whose skewnesses fall outside this range, estimat
the kurtoses of surviving samples and find that 98.73% of these lie between 1.932 anc
4.797. Eliminating those samples whose kurtoses lie outside this range, we have eliminate
altogether $(1 - 0.01275)^4 = 5\%$ of the samples with which we started. Given that we ar
operating under the null hypothesis of a standard normal, this rejection rate correspond
to our type I error rate or significance level. More generally, if we reject a proportion c
at each stage, we would get an overall significance level β equal to $1 - (1 - \alpha)^4$.[7]

Once we have the bounds, the implementation of the test is straightforward. We star
with a sample of P/L observations. We run these through the two transformations 4.1 anc
4.2 to obtain our z series, and we estimate its first four sample moments. We then choose
our (overall) significance level β and estimate the bounds of the first, second, third anc
fourth moment tests using Monte Carlo simulation as just described. The null hypothesi
(of model adequacy) is then accepted if each sample moment lies between these bound
and is rejected otherwise. For example, if $n = 50$, we 'pass' the model if the sample mear
lies in the range -0.350 to 0.351, if the sample variance lies in the range 0.754 to 1.256
if the sample skewness lies in the range -0.850 to 0.847 and if the sample kurtosis lie
in the range 1.932 to 4.797. The model 'fails' if any of these conditions is not met.[8]

Some estimates of the different moment test bounds are given in Table 4.2. These ar
obtained using $50\,000$ simulation trials for $\alpha = 0.01275$ corresponding to an overal

Table 4.2 Bounds of individual moment tests[1]

Moment number	$n = 50$		$n = 100$		$n = 250$		$n = 500$		$n = 1000$	
	Lower	Upper	Lower	Upper	Lower	Upper	Lower	Upper	Lower	Upp ϵ
1	-0.350	0.351	-0.251	0.250	-0.159	0.159	-0.111	0.112	-0.079	0.07
2	0.754	1.256	0.826	1.180	0.889	1.113	0.921	1.079	0.945	1.05
3	-0.850	0.847	-0.614	0.614	-0.388	0.388	-0.272	0.274	-0.191	0.19
4	1.932	4.797	2.155	4.394	2.399	3.924	2.545	3.621	2.657	3.43

Notes:
1. The significance level for each individual moment test is 0.0127, making for an overall rejection rate of 0.0
after all moment tests have been carried out. Tests are based on $50\,000$ simulation trials.

significance level of $\beta = 0.05$. This table shows the bounds for each individual moment test statistic for the same n-values as before (i.e. $n = 50, 100, 250, 500$ and 1000). As we would expect, these bounds narrow as n gets bigger. For example, if we take the first moment test, the bounds range -0.350 to 0.351 for a sample size of $n = 50$ narrows to -0.251 to 0.250 for $n = 100$ and so forth.

This testing framework is calibrated to ensure that each individual moment test gives the 'correct' proportion of failures under the null given that we wish each individual moment prediction to have equal importance in the rejection process. This means that the rejection rates are exactly what we want. The only source of estimation error is in the bounds obtained, but if we want to increase the accuracy of our bound estimates, we can do so by increasing the number of simulation trials m: the accuracy will increase with the square root of m.

Table 4.3 shows the power results for the moments test applied with an overall significance level β equal to 0.05. These results show that the moments test has considerable power against most of the alternative hypothesis distributions considered. The power of our test also rises appreciably with the sample size. For example, in the last line of the table, where we have a skewed leptokurtic alternative distribution, the power of the test rises from 0.2245 for $n = 50$ to 0.9998 for $n = 1000$. The test also tends to have greater power against more pronounced departures from standard normality. Evidence of this can be seen by comparing, e.g. the fourth and last alternatives: the former is quite close to standard normality, and the latter is much less so, and our test has much more power against the latter than the former.

Comparing the LR and moments tests, we can say that the moments test is a little less powerful than the LR test in the presence of departures from H_0^a, but the difference in power is usually small in absolute terms. For example, with $n = 1000$, the average power of the LR test against the H_0^a alternatives is 0.927, whereas that of the moments test is 0.903. However, the moments test is very much superior to the LR test in the face of departures from H_0^b, and its superiority tends to increase as the sample size gets bigger. For $n = 50$, the average power of the LR test against the non-normal alternatives considered in the table is 0.082, and this rises to 0.092 for $n = 1000$. But for the moments test, the average power rises from 0.0181 for $n = 50$ to 0.733 for $n = 1000$. The power of the LR barely improves as n gets bigger, but the power of the moments test improves markedly.

Thus, in choosing the LR test, we get a slightly superior test against departures from H_0^a at the price of very little power in the face of departures from H_0^b. On the other hand, in choosing the moments test, we get a vastly improved ability to detect departures against H_0^b at the price of a small loss of power in the face of departures from H_0^a. In other words, the moments test is much more robust than the LR test, in the sense that it is able to detect a much wider range of departures from the null hypothesis.

5. An illustration

It might help to give a 'walked through' example of how the test would be implemented. It is also helpful to choose an example that illustrates the ability of the proposed test to identify a common source of model inadequacy – namely, the under-estimation of kurtosis. This is a common problem in financial risk modelling but is also an important one because under-estimation of kurtosis leads to under-estimation of VaR and other risk

Table 4.3 Power results for moments test[1]

Departure from standard normality	Type of distribution	Mean	Std	Skew	Kurtosis	Power for sample size (n)				
						$n = 50$	$n = 100$	$n = 250$	$n = 500$	$n = 1000$
Alternative involves normal with non-standard parameter values										
Mean $\neq$ 0	Non-standard normal	0.1	1	0	3	0.0809	0.0976	0.2203	0.4178	0.7532
Std $\neq$ 1	Non-standard normal	0	1.1	0	3	0.1186	0.1844	0.4207	0.7249	0.9656
Mean $\neq$ 0 and std $\neq$ 1	Non-standard normal	0.1	1.1	0	3	0.1495	0.2343	0.5394	0.8303	0.9904
Alternative involves $\mu = 0$, $\sigma = 1$ and non-normal distribution										
Skewed	Skew-t with $\nu = 100$ and $\lambda = 25^2$	0	1	0.11	≈ 3	0.0630	0.0722	0.0996	0.1575	0.2831
Skewed	2PN with $\sigma_2 = 1.1^3$	0	1	0.16	≈ 3	0.0549	0.0689	0.0967	0.1847	0.3595
Skewed	2PN with $\sigma_2 = 1.2^3$	0	1	0.32	≈ 3	0.0825	0.1399	0.3469	0.7005	0.9679
Leptokurtic	Error distribution with $c = 1.25^4$	0	1	0	3.56	0.1066	0.1340	0.2221	0.3676	0.6554
Leptokurtic	Generalized symmetric-t with $\nu = 15^5$	0	1	0	3.55	0.1132	0.1445	0.2215	0.3560	0.5967
Leptokurtic	Generalized symmetric-t with $\nu = 5^5$	0	1	0	9	0.3702	0.5667	0.8803	0.9923	1.0000
Skewed and leptokurtic	Skew-t with $\nu = 5$ and $\lambda = 3^2$	0	1	1	3.77	0.4300	0.8835	0.9996	1.0000	1.0000
Skewed and leptokurtic	Skew-t with $\nu = 15$ and $\lambda = 5^2$	0	1	0.61	4.12	0.2245	0.4082	0.7787	0.9694	0.9998

Notes:
1. Significance level (type I error rate) = 0.05, and each result is based on the outcome of 10 000 Monte Carlo simulation trials.
2. The skew-t parameters are those of the (ν, λ) parameterization in Jones and Faddy (2003), with mean and standard deviation set to 0 and 1, respectively.
3. The 2PN parameters are those of the $(\mu, \sigma_1, \sigma_2)$ parameterization of John (1982), with mean and standard deviation set to 0 and 1, respectively.
4. The error distribution parameters are as given in Evans *et al.* (2000), with mean and standard deviation set to 0 and 1, respectively.
5. The symmetric-t parameters are as given in Evans *et al.* (2000), with mean and standard deviation set to 0 and 1, respectively.

measures at high-confidence levels and means that institutions are more exposed to the risk of high losses than their models suggest. This underestimation will manifest itself in z coming from a heavy-tailed distribution instead of a normal one.[9] Let us suppose, therefore, that our z-observations are drawn from a generalized t distribution with mean 0, standard deviation 1 and 5 degrees of freedom. This distribution has a zero skew and a kurtosis equal to 9, whereas a standard normal has a zero skew and a kurtosis equal to 3.

Now suppose that we have a sample of 250 z observations drawn from this distribution. We begin by estimating the sample mean (-0.0316) and comparing it to the bounds for the mean test [which are (-0.159, 0.159); see Table 4.3]. As the sample mean is well within this range, the model therefore passes the first test. We then compare the sample standard deviation (0.9895) with its predicted range (0.889 to 1.113), and find that the model passes the second test as well. We now proceed to the skewness test: the sample skewness (0.1394) is inside its predicted range (-0.388 to 0.388), so the model passes this test too. We then carry out the kurtosis test: the sample kurtosis (4.4797) is outside its predicted range (2.399–3.924). The model therefore fails the kurtosis test, and so fails overall. By comparison, if we applied the LR test, we would find that the model would pass. This is as we would expect, because the model inadequacy lies in one of the higher moments that the LR test does not address.

This example is only illustrative because it is based on a single sample, but it is plausible given the power results reported in Tables 4.1 and 4.3. In particular, the power of the LR test against this type of alternative hypothesis is only 0.206 (see Table 4.1, third line from bottom), whereas the power of the moments test in this situation is 0.8803 (Table 4.3, third line from bottom). We can therefore say with some confidence that the LR test is likely to give the wrong answer in such circumstances, whereas the moments test is likely to give the right one.

This example also shows how easy it is to apply the test in practice: all implementation requires are the sample moments of z and knowledge of the relevant risk bounds. The latter can be estimated by simulation in the course of a testing exercise, but in practice, it is much easier if these are pre-estimated and then distributed to users. In principle, this needs doing only once, and once obtained, they can be tabulated for others to use. There is therefore no need for a model evaluator as such to estimate these bounds or even have the skills to do so. Instead, the evaluator only needs to look up these values and compare them to the sample moments that he or she has just estimated. Implementation of the model on a spreadsheet is then very straightforward.[10]

6. Conclusions

This chapter examines the important practical problem of testing the adequacy of a risk forecasting model. Its point of departure is the LR test applied to the PIT-then-Berkowitz transformed series z, which should be standard normal under the null hypothesis of model adequacy. Applied to z, the LR test has good power against normal departures from standard normality but poor power against non-normal departures: the LR test is poor at detecting skewness and excess kurtosis in z, and these are common risk management problems. To remedy this defect, the chapter proposes a new procedure that combines tests of all first four moment predictions into a single test. The proposed test can be regarded as a robust omnibus test of risk model adequacy and gives equal weight to each moment prediction, which is desirable if one has no a priori sense of which moment prediction might be problematic in any given situation. Simulation results suggest that this test has considerable power against a wide range of departures from standard normality and is consequently much more robust than the LR test in the face of possible model misspecification. The test is also very easy to implement and does not require any statistical or econometric sophistication on the part of the model evaluator. Indeed, the test can easily be pre-programmed into a spreadsheet, and all the evaluator then needs to do is input the data and interpret the test result.[11]

With this use in mind, Table 4.4 provides estimates of the bounds for various sample sizes that users are likely to encounter in practice. This table enables users to implement the test without having to simulate the bounds themselves.

Table 4.4 Test bounds for various sample sizes[1]

n^2	Mean		Std		Skewness		Kurtosis	
	Lower	Upper	Lower	Upper	Lower	Upper	Lower	Upper
50	−0.350	0.351	0.754	1.256	−0.850	0.847	1.932	4.797
75	−0.288	0.288	0.800	1.204	−0.710	0.700	2.059	4.554
100	−0.251	0.250	0.826	1.180	−0.614	0.614	2.155	4.394
125	−0.223	0.225	0.844	1.161	−0.551	0.550	2.216	4.299
150	−0.202	0.205	0.859	1.145	−0.503	0.502	2.266	4.181
175	−0.189	0.190	0.867	1.135	−0.469	0.470	2.303	4.087
200	−0.178	0.177	0.876	1.126	−0.436	0.436	2.339	4.024
225	−0.169	0.167	0.882	1.119	−0.412	0.413	2.371	3.964
250	−0.159	0.159	0.889	1.113	−0.388	0.388	2.399	3.924
275	−0.151	0.150	0.894	1.107	−0.372	0.367	2.422	3.868
300	−0.145	0.145	0.898	1.102	−0.356	0.348	2.440	3.827
325	−0.140	0.140	0.901	1.099	−0.341	0.335	2.459	3.791
350	−0.134	0.135	0.905	1.095	−0.332	0.322	2.475	3.772
375	−0.128	0.129	0.909	1.092	−0.316	0.312	2.490	3.741
400	−0.124	0.124	0.912	1.089	−0.306	0.303	2.505	3.708
425	−0.121	0.121	0.914	1.087	−0.297	0.293	2.517	3.681
450	−0.117	0.118	0.917	1.083	−0.286	0.288	2.527	3.659
475	−0.115	0.115	0.919	1.081	−0.277	0.279	2.537	3.641
500	−0.111	0.112	0.921	1.079	−0.272	0.274	2.545	3.621
525	−0.109	0.109	0.923	1.077	−0.265	0.266	2.555	3.607
550	−0.106	0.107	0.925	1.076	−0.260	0.261	2.563	3.598
575	−0.103	0.106	0.926	1.074	−0.255	0.253	2.571	3.588
600	−0.101	0.103	0.929	1.073	−0.248	0.245	2.577	3.570
625	−0.099	0.101	0.930	1.071	−0.241	0.241	2.585	3.558
650	−0.097	0.099	0.931	1.070	−0.238	0.235	2.591	3.549
675	−0.094	0.096	0.932	1.068	−0.234	0.231	2.596	3.546
700	−0.093	0.095	0.934	1.067	−0.231	0.229	2.603	3.532
725	−0.092	0.093	0.935	1.066	−0.226	0.223	2.608	3.523
750	−0.091	0.091	0.936	1.065	−0.222	0.219	2.611	3.514
775	−0.090	0.090	0.937	1.063	−0.217	0.216	2.615	3.505

Table 4.4 Continued

n^2	Mean		Std		Skewness		Kurtosis	
	Lower	Upper	Lower	Upper	Lower	Upper	Lower	Upper
800	−0.089	0.088	0.938	1.063	−0.215	0.213	2.620	3.497
825	−0.087	0.087	0.939	1.062	−0.210	0.209	2.630	3.482
850	−0.086	0.086	0.940	1.061	−0.207	0.207	2.636	3.477
875	−0.086	0.084	0.941	1.060	−0.202	0.204	2.637	3.470
900	−0.084	0.083	0.942	1.059	−0.199	0.201	2.641	3.461
925	−0.083	0.082	0.943	1.058	−0.198	0.197	2.646	3.456
950	−0.081	0.081	0.943	1.058	−0.195	0.196	2.649	3.449
975	−0.080	0.080	0.944	1.057	−0.193	0.194	2.654	3.443
1000	−0.079	0.079	0.945	1.057	−0.191	0.190	2.657	3.438

Notes:
1. The significance level for each individual moment test is 0.0127, making for an overall rejection rate of 0.05 after all moment tests have been carried out. Tests are based on 50 000 simulation trials.
2. For intermediate values of n, approximate bounds can be obtained using interpolation.

7. Acknowledgements

This work was carried out under an ESRC research fellowship on 'Risk Measurement in Financial Institutions' (RES-000-27-0014), and the author thanks the ESRC for their financial support. He also thanks Jeremy Berkowitz and Ken Wallis for helpful feedback. However, the usual caveat applies.

References

Aït-Sahalia, Y. and Lo, A. (1998) Nonparametric estimation of state-price densities implicit in financial asset prices. *Journal of Finance*, **53**, 499–547.
Bao, Y., Lee, T.-H. and Saltoglu. B. (2004) A test for density forecast comparison with applications to risk management. Mimeo, University of California, Riverside, and Istanbul, Marmora University.
Berkowitz, J. (2001) Testing density forecasts, with applications to risk management. *Journal of Business and Economic Statistics*, **19**, 465–74.
Bontemps, C. and Meddahi, N. (2003) Testing normality a GMM approach. Mimeo, CIRANO, University of Montreal.
Chatfield, C. (1993) Evaluating interval forecasts. *Journal of Business and Economic Statistics*, **11**, 121–35.
Christoffersen, P.F. (1998) Evaluating interval forecasts. *International Economic Review*, **39**, 841–62.
Corradi, V. and Swanson, N.R. (2003) *Bootstrap Specification Tests for Diffusion Processes*. Mimeo: Queen Mary College and Rutgers University.
Crnkovic, C. and Drachman, J. (1995) Quality control. *Risk*, **9** (September), 139–43.
Diebold, F.X., Gunther, T.A., and Tay, A.S. (1998) Evaluating density forecasts with applications to financial risk management. *International Economic Review*, **39**, 863–83.
Dowd, K. (2004) A modified Berkowitz back-test. *Risk*, **17** (April), 86.
Duan, J.-C. (2003) *A Specification Test for Time Series Models by a Normality Transformation*. Mimeo: Rotman School of Management, University of Toronto.

Evans, M., Hastings, N., and Peacock, B. (2000) *Statistical Distributions*. Third edition. New York: John Wiley and Sons.

Hong, Y. and Li, H. (2002) *Nonparametric Specification Testing for Continuous-Time Models with an Application to Spot Interest Rates*. Mimeo, Cornell University.

John, S. (1982) The three-parameter two-piece normal family of distributions and its fitting. *Communications in Statistics – Theory and Methods*, **11**, 879–85.

Jones, M.C. and Faddy, M.J. (2003) A skew extension of the *t*-distribution, with applications. *Journal of the Royal Statistical Society B*, **65** (Part 1), 159–74.

Kupiec, P. (1995) Techniques for verifying the accuracy of risk management models. *Journal of Derivatives*, **3**, 73–84.

Wallis, K.F. (2004) An assessment of Bank of England and National Institute inflation forecast uncertainties. *National Institute Economic Review*, No. 189, July.

Notes

1. However, there are many other circumstances where we might wish to evaluate a model that is not known but merely estimated. This is usually the case in volatility forecasting or interest-rate modelling. The evaluation of estimated models is a more difficult problem and is the subject of a parallel literature (Hong and Li, 2002; Bontemps and Meddahi, 2003; Corradi and Swanson, 2003; Duan, 2003; Bao et al. 2004). The extent to which the approach suggested here will apply to estimated density forecast models is therefore an open question although it is reassuring to note that studies that have looked at parameter uncertainty seem to conclude that it is (usually?) of second-order importance [e.g. Chatfield (1993) or Diebold et al. (1998)].

2. If the risk model is adequate and if the forecast horizon matches the frequency with which the random variable is observed, then z should be iid standard normal, and not just plain standard normal. For example, if we observe P/L on a daily basis and our model forecasts a daily VaR, then z should be iid standard normal. It would also be the case that the z should be iid if there is a multi-day holding period, but the multi-day VaR was obtained by extrapolating a daily VaR forecast over a longer period. This paper does not address how we might test the iid prediction, where it arises, but Berkowitz (2001, pp. 468–469) has a good discussion of how to test for iid standard normality within an LR context. For example, he suggests we might estimate a first-order autoregressive process for z and then test whether the autoregression parameter is significantly different from 0.

3. More specifically, the first line gives results for the case where the 'true' distribution is a normal with mean 0.1 and standard deviation 1, and the second and third lines give corresponding results for normal distributions with $\mu = 0, \sigma = 1.1$ and $\mu = 0.1, \sigma = 1.1$, respectively. Because of their normality, these all have zero skewness and kurtosis equal to 3. The fourth to eleventh lines then give standardized ($\mu = 0, \sigma = 1$) results for a skew-*t* with skew 0.11 and kurtosis (about) 3; a 2PN with skew 0.16 and kurtosis (about) 3; a 2PN with skew 0.32 and kurtosis (about) 3; an error distribution with skew 0 and kurtosis 3.56; a symmetric-*t* with zero skew and kurtosis 3.55; a symmetric-*t* with zero skew and kurtosis 9; a skew-*t* with skew 1 and kurtosis 3.77 and a skew-*t* with skew 0.61 and kurtosis 4.17. More details on these distributions are given in the Appendix.

4. This chapter is not the first to suggest looking at the higher moments in a z-testing framework. Berkowitz himself suggested that lagged and/or powered values of z, including powers of 3 or more, could be incorporated into his regression framework and therefore taken account of [Berkowitz (2001, p. 468)]. Berkowitz was therefore aware of the higher moment issue on which the present paper expands but did not discuss the issue in detail or present any power results that explicitly take account of higher moment departures from standard normality. However, his suggestion that higher moment departures can be accounted for within an LR testing framework is well worth further exploration.

5. H_0^b also makes predictions about all the higher moments – the fifth moment should be 0, etc. – but we ignore these for the moment as risk managers are usually concerned only with the first four moments. However, the testing framework suggested here can easily be modified to handle the higher moment predictions as well. See Note 8 below.

6. As an aside, we also want a testing framework that does not make unwarranted assumptions about how individual moment test statistics might be related. In particular, we cannot (and do not) assume that the moments are independent of each other.
7. In practice, we would start with the overall significance level β and then obtain the associated α-value using the relationship $\alpha = 1 - (1 - \beta)^{1/4}$.
8. It makes sense to focus on the first four moment predictions because these are the ones with which risk managers are mainly concerned. However, if we wish to, we could easily modify the testing framework to handle higher moments. For example, if we wanted to incorporate the fifth moment prediction and have an overall significance level of β, we would choose $\alpha = 1 - (1 - \beta)^{1/5}$. We would then add a fifth moment test comparable to the individual four moment tests considered in the text and calibrate accordingly. Once we have the five sets of bound estimates, one for each moment, we would apply the actual test. This would involve estimating the first five moments of our z-sample, and we would reject the model if any sample moment falls outside the relevant bounds. Extensions to incorporate sixth and higher moment predictions are obvious.
9. A more detailed illustration of this point is provided by Dowd (2004, p. 86).
10. This ease of use can be a major consideration. For example, the Basel capital adequacy regulations make banks' regulatory supervisors responsible for validating bank's risk models. However, these supervisors are typically qualified as accountants and are rarely highly trained in statistics. The simplicity of the framework proposed is therefore a major benefit to such users.
11. The proposed testing framework is also no doubt capable of refinement, and it may be possible to improve the power of the omnibus test if we replace the basic moment tests used here with alternatives. However, the chances are that any such improvements in power would come at the price of making the tests less user-friendly, and (as the previous footnote points out) this is certainly an issue in the risk management context.

Appendix

This appendix gives details of the error distribution, 2PN distributions, t distributions and skew-t distributions used as alternative hypotheses to generate some of the simulation results reported in the tables.

1. Error distribution

The error distribution uses three parameters – a location parameter a, equal to the mean; a scale parameter b and a shape parameter c – and its pdf is

$$p(x) = \frac{\exp[-(|x - a|/b)^{2/c}/2]}{b2^{c/2+1}\Gamma(1 + c/2)}, \tag{4.2}$$

where $\Gamma(.)$ is a gamma distribution.

Its variance is equal to $2^c b^2 \Gamma(3c/2)/\Gamma(c/2)$, and the normal is a special case where $c = 1$.

The error distribution used in the simulations is parameterized so that the mean is 0, the standard deviation is 1 and $c = 1.25$. It therefore has a zero skew and a kurtosis of 3.56.

2. Two-piece normal distribution

The 2PN distribution has three parameters – a mode μ and two uncertainty parameters σ_1 and σ_2 – and its pdf is

$$
p(x) = \begin{cases} C \exp\left\{-\dfrac{1}{2\sigma_1^2}(x-\mu)^2\right\}, x \le \mu \\[2mm] C \exp\left\{-\dfrac{1}{2\sigma_2^2}(x-\mu)^2\right\}, x \ge \mu \end{cases},
\tag{4.3}
$$

where $C = k(\sigma_1 + \sigma_2)^{-1}$ and $k = \sqrt{2/\pi}$ (John, 1982). The distribution takes the lower half of a normal distribution with parameters μ and σ_1 and the upper half of a normal with parameters μ and σ_2, scaled using scaling factors $2\sigma_1/(\sigma_1 + \sigma_2)$ and $2\sigma_2/(\sigma_1 + \sigma_2)$. The distribution is negatively (positively) skewed if $\sigma_1 > \sigma_2 (\sigma_1 < \sigma_2)$. If $\sigma_1 = \sigma_2$, then the 2PN is symmetric and equivalent to a normal with mean/mode μ and standard deviation σ_1 (or σ_2).

The mean and variance of the 2PN are $\mu + \sqrt{2/\pi}(\sigma_2 - \sigma_1)$ and $(1 - 2/\pi)(\sigma_2 - \sigma_1)^2 + \sigma_1\sigma_2$ (Wallis, 2004).

The 2PN distributions used in the simulations are calibrated so that the mean and standard deviation are equal to 0 and 1, and σ_2 takes the values 1.1 and 1.2

3. t-Distribution

The t-distributions used in the simulations are conventional symmetric generalized t distributions (i.e. not Student-t distributions) with mean 0, standard deviation 1, and 5 and 15 degrees of freedom. They both therefore have zero skew, but the former has a kurtosis of 9 and the latter has a kurtosis of 3.55.

4. Skew-t distribution

There are various parameterizations of skew-t distributions, but the one used in this chapter is the (v, λ) parameterization suggested by Jones and Faddy (2003). Its pdf is:

$$
p(x) = C_{v,\lambda}^{-1}\left[1 + \frac{x}{(v+x^2)^{1/2}}\right]^{\frac{v+\lambda+1}{2}}\left[1 - \frac{x}{(v+x^2)^{1/2}}\right]^{\frac{v-\lambda+1}{2}}
\tag{4.4}
$$

where $C_{v,\lambda}^{-1} = 2^{v-1}B\left(\frac{v+\lambda}{2}, \frac{v-\lambda}{2}\right)$ and $B(.)$ is a beta distribution. v is closely related to the degrees of freedom, and λ indicates possible skewness. The skew-t takes the symmetric-t as a special case where $\lambda = 0$, and it takes the normal as a special case where $\lambda = 0$ and $v \to \infty$.

The skew-t distributions used to produce the simulation results reported in this paper are standardized to have zero mean and unit standard deviation. The values of (v, λ) used are (100,25), (5,3) and (15,5).

5 Measuring concentration risk in credit portfolios

*Klaus Duellmann**

Abstract

This paper addresses the need to measure the concentration risk in credit portfolios, which is not captured by banks' minimum capital requirements in the internal rating-based approaches under Basel II. Concentration risk can arise from an unbalanced distribution of exposures to single borrowers (single name concentrations) or sectors (sectoral concentrations), the latter being more difficult to discern and address using quantitative models. Simple concentration indices, an extension of the single-factor model on which the regulatory minimum capital requirements are based and more complex multi-factor portfolio models are presented as tools to quantify these risks. A case study-like numerical example demonstrates their use and confirms their relative importance for the measurement of portfolio credit risk.

1. Concentration risk and validation

An important innovation in the internal ratings-based (IRB) approach of Basel II is the introduction of model-based risk weight functions to determine the risk-weighted assets for credit risk in the banking book. As the IRB risk weight functions are already given by the new Framework, the internal validation process does not focus on the underlying *IRB model* itself but instead on its input variables, namely the probability of default (PD), the loss given default (LGD) and the exposure at default (EAD). More precisely, validation in this context 'encompasses a range of processes and activities that contribute to an assessment of whether ratings adequately differentiate risk, and whether estimates of risk components (such as PD, LGD, or EAD) appropriately characterise the relevant aspects of risk'.[1] Whereas validation defined in this way focuses on the inputs of the IRB model, this paper is concerned with the model's output. More specifically, it addresses the impact of risk concentrations in credit portfolios on capital in relation to the IRB model and compares selected methods how to quantify this impact.

Risk concentrations are defined by credit risk that arises from portfolio concentrations in single borrowers (*name concentration* or *low granularity*) and in geographical or business sectors (*sectoral concentration* or *segment concentration*). A quantification of concentration risk is related to the validation of IRB systems in a broad sense in that the presented methods are intended to help explain differences between Pillar 1 minimum

---

* Deutsche Bundesbank, Frankfurt, Germany. The views expressed here are my own and do not necessarily reflect those of the Deutsche Bundesbank.

capital requirements and capital estimates from an internal model. It turns out that the IRB risk weight functions can understate credit risk in portfolios in which the assumptions of the IRB model are not fulfilled, in particular if the portfolio is concentrated in borrower names, business sectors or geographical regions. A quantitative assessment of the potential consequences in terms of capital for such portfolios is therefore warranted. This paper discusses various approximation methods that could be useful for this purpose and that avoid the computational burden of Monte Carlo simulations.

As risk concentrations are not fully captured by the regulatory minimum capital requirements (Pillar 1), a treatment under Pillar 2 of the new Framework, the supervisory review process, is warranted. The Basel II framework text states that 'Banks should explicitly consider the extent of their credit risk concentrations in their assessment of capital adequacy'.[2] This issue becomes all the more important as the scope of IRB banks is not limited to large, internationally active banks – regional and local banks, for example in the European Union, may also be eligible to use IRB approaches.

From a general perspective, concentration risk can affect risk management in various places of the enterprise. 'Risk concentrations can arise in a bank's assets, liabilities, or off-balance sheet items, through the execution or processing of transactions (either product or service), or through a combination of exposures across these broad categories'.[3] In the trading book, counterparty risk is also a case in point. Consider, for example, the global credit derivatives market, where the top ten counterparties are responsible for around 85% of the total market volume. Therefore, concentration risk can have many facets in banking practice. In the following, I will focus only on risk concentration in credit portfolios.

2. Concentration risk and the IRB model

An alternative interpretation of concentration risk is to understand it as the flip side of *diversification*. Intuitively, the better a credit portfolio is diversified across names or sectors, the lower its 'concentration risk' should be. It has been argued that the IRB model does not capture diversification effects which could be misunderstood as if real bank portfolios generally warranted capital relief, depending on how far they are diversified. As will become clearer in the following, the IRB model is based on strict assumptions concerning the risk concentration in the credit portfolio. These assumptions imply that Pillar 1 does not necessarily produce a conservative capital estimate, for example if a portfolio is concentrated in names or sectors.

The importance of exposure concentrations in credit portfolios is recognized in the Basel II framework text: 'Because lending is the primary activity of most banks, credit risk concentrations are often the most material risk concentrations within a bank'.[4] This statement is confirmed by the recent empirical work of Duellmann and Masschelein (2006), based on central credit register data. They find that unbalanced exposure distributions across sectors in real banks can justify up to 40% higher capital cushions than credit portfolios, which resemble the aggregate credit exposure distribution across sectors of the German banking system as a whole. The impact of low granularity appears to be smaller, as realistic portfolios can require around 8% higher capital than a very fine-grained portfolio. This number is also similar to the empirical results in Gordy and Lütkebohmert (2007). The empirical results on the materiality of sector concentration and granularity are also broadly in line with Heitfield *et al.* (2006), who use syndicated loan portfolios in the US. In summary, these empirical findings indicate that both sectoral concentration and name concentration can have a material influence on the value-at-risk (VaR), with sectoral concentrations tending to have the stronger impact.

Measuring the impact of risk concentrations in terms of capital as is done, for example in the above-cited empirical studies, requires a credit risk model. A key issue in this case is how to capture dependencies of credit events. Credit events can be either borrower defaults in the world of book-value accounting or adverse changes in the credit quality of the borrower from the perspective of a mark-to-market valuation of credit claims. The models discussed in the following are all one-period default-mode models. The IRB model, although technically a default-mode model, was calibrated to a mark-to-market framework by adding a maturity adjustment term as a multiplicative factor. I will point out the implications for the various methodologies as I go along.

To measure name concentration, all exposures to the same borrower need to be aggregated and borrowers in the same sector likewise need to be aggregated to quantify sectoral concentration. From this perspective, name concentration and sectoral concentration appear to differ only in the aggregation level. But this view ignores a key difference between them, which determines the methodological differences in the ways how they are measured.

If the credit quality of a single borrower deteriorates for firm-specific reasons, this generally does not affect the individual credit risk of other borrowers in the portfolio. In other words, name concentration represents idiosyncratic risk that can generally be diversified away by increasing the number of exposures. Therefore, exposure size is a key determinant of this type of concentration risk, and simple ad hoc measures of the single-exposure distribution in a credit portfolio can already provide useful information.

Risk concentrations in sectors instead require technically more sophisticated methods for the following reason. A deterioration in credit quality of borrowers in the same sector is not an idiosyncratic event but caused by the same sector affiliation, in economic terms by the dependence of these borrowers' credit quality on a common sector-specific risk factor. Therefore, measuring sectoral concentration requires the modelling of credit dependencies, typically by a factor structure that affects the stochastic trigger variable of a borrower default, for example, a firm's asset value returns. As a consequence, measuring credit risk from sectoral concentrations is more challenging from a methodological perspective, as the required risk parameters – for example asset correlations – are more difficult to estimate than for name concentration.

The distinction between name concentration and sectoral concentration is particularly convenient in the context of the IRB model, but it is less often made explicit in common multi-factor models. A conceptual reason is that the risk contribution – more precisely, the marginal VaR – of an individual exposure in a multi-factor model implicitly captures both aspects of concentration risk. It also explains why concentration risk is generally not regarded as a separate risk category but is instead woven into the credit risk of the portfolio. In cases in which the credit risk model cannot (fully) account for both types of concentration risk, it is, however, important to measure them separately to validate the capital figure produced by the model. The IRB model is a case in point, which is illustrated in the rest of this section in more detail.

Under the Basel I regime, the required regulatory minimum capital for credit risk in the banking book depends only on the borrower type and does not explicitly account for dependencies between credit events. One of the main goals of Basel II is to increase risk sensitivity and to better align regulatory minimum capital requirements with the economic capital determined by the bank for internal risk management purposes. This is achieved by replacing the rigid, borrower-dependent risk weight scheme of Basel I with risk weight functions that are based on a credit risk model and capture dependencies of credit events through asset correlations.

The strong sensitivity of the capital output to the factor structure and the empirical difficulties in obtaining robust estimates of asset correlations may be factors that contributed to regulators' decision not to allow banks to use their internal portfolio models for the calculation of regulatory minimum capital in Basel II. By contrast, the IRB risk weight functions are derived from a given model that captures systematic risk by a common dependence of the credit quality of different borrowers on a single, unobservable risk factor. The asset correlations in Pillar 1 depend on the asset class, in most asset classes also on the borrower's PD and for loans to small- and medium-sized enterprises (SMEs) also on firm size.

An important advantage of selecting a *single risk factor model* for the calculation of regulatory capital requirements is a property called *portfolio invariance*. By this property, risk-weighted assets can be calculated on an exposure-by-exposure basis without considering the risk characteristics of the remaining portfolio. In credit risk models, the contribution of individual exposures to the total portfolio risk generally depends on the risk characteristics of the other exposures; yet, Gordy (2003) showed that portfolio invariance can hold even in heterogeneous credit portfolios – at least asymptotically – under the following two assumptions:[5]

The first assumption requires that the portfolio is extremely fine-grained in terms of exposures to individual borrowers. Loosely speaking, the contribution of every borrower's loan exposure to the total portfolio risk has to be very small. This assumption is often referred to as *infinite granularity*. It implies the absence of *name concentration*, which describes risk from large exposures to single borrowers. In the case of the IRB model, it is not only the existence of large exposure lumps but in fact any deviation from an infinitely fine-grained portfolio, which violates this model assumption. Clearly, from an economic perspective a more important factor than the violation of the assumption as such is the materiality of its impact on capital. It turns out that name concentration is generally less important for large, well-diversified banks but can be more relevant for smaller, regional banks. An empirical S&P study for the 100 largest rated western European banks uncovered still significant credit concentrations in many of the relatively large banks in the sample.[6]

The second assumption of the IRB model requires that systematic risk be fully captured by a *single* systematic risk factor. Clearly, credit portfolios with exposures in the same regions and industries are affected by sector-specific common risk factors. But, for a bank that is well-diversified across regions and industry sectors, it is conceivable that all systematic risk not captured by a single common factor, such as a global economic cycle, is eliminated. A single-factor model can, however, produce a capital figure that is too low if a credit portfolio is concentrated in certain sectors.

In summary, the two assumptions of the IRB model needed for portfolio invariance are closely related to the two types of risk concentration in the portfolio. As the quantification of credit risks from name concentrations and sector concentrations requires quite different methodologies, these types of risk concentrations are treated separately in the next two sections.

3. Measuring name concentration

For any quantification of risks, it is convenient to have quantitative benchmarks, for example to measure the distance from a neutral reference state of no concentration or full diversification. The common but not very precise definition of name concentration as referring to a lumpy exposure distribution across borrowers makes it difficult to define such a benchmark. If name concentration is defined instead as an aberration from the IRB model, the special case of an infinitely granular portfolio lends itself as a natural benchmark for the quantification of name concentration. This basic idea of measuring name concentration as the capital impact of not holding a fully diversified portfolio has also been suggested in the industry.[7] Note that this definition, which is more convenient for the purpose of this paper, implies that concentration risk is already present in credit portfolios that generally would not be regarded as particularly lumpy. In such cases, the risk contribution from exposure concentrations can be very small and negligible in practice.

Measuring the impact of name concentrations on portfolio risk requires robust quantitative tools that can be either ad hoc or model-based methods. A prominent example of an ad hoc method is the Herfindahl-Hirschman Index (HHI). It has found wide application in various areas of economic theory, one example being the measurement of competition in a market. For the purpose of measuring name concentration, the HHI is defined as the sum of all N-squared relative portfolio shares w_n of the exposures to single borrowers: $HHI = \sum_{n=1}^{N} w_n^2$. By construction, the HHI is restricted between zero and one, with one corresponding to monopolistic or full concentration. Practitioners also employ the *Herfindahl number*, defined as the inverse of the HHI.

Key advantages of the HHI include its simplicity and the relatively moderate data requirements, i.e. only the exposure size per borrower. Both advantages come at a price. The HHI can be misleading as a risk measure because of two limitations. Firstly, it does not consider the borrower's credit quality, and secondly, it does not account for credit risk dependencies between borrowers. An exposure to a Aaa-rated borrower, for example, is treated in the same way as an exposure to a B-rated borrower. Two large exposures to borrowers belonging to the same supply chain and located in the same town are treated in the same way as two large exposures to borrowers in completely unrelated industry sectors and located on different continents.

The first limitation can be addressed by a 'rating-weighted' HHI. The squared relative exposure shares are weighted in the aggregation by a numeric borrower rating, thereby giving more weight to borrowers with a lower credit quality. The second limitation, the ignorance of default dependencies, cannot be remedied in a similarly elegant way. As a rating-weighted HHI provides a ranking in terms of name concentration, it can be useful for a comparison across credit portfolios; yet, it fails to produce a capital figure for low granularity

Although the data requirements for the HHI calculation seem to be moderate at first sight, they often turn out to be difficult to meet in practice. They require nothing less than the aggregation of all exposures to the same borrower for the whole business entity, be it a bank or a banking group. A heterogeneous IT environment can already present a serious technical obstacle to this aggregation. Furthermore, large borrowers, which are actually the most relevant entities for measuring name concentration, can themselves be complex structures that are not always easy to identify as belonging to the same borrower entity.

The fast-growing markets for securitization and credit derivatives offer new opportunities to mitigate concentration risk. It can, however, be difficult to identify single borrowers for example in large asset pools, and to quantify the real exposure, such as in complex basket products. Securitization structures, in which the credit risk of an exposure to a specific borrower depends on how the credit risk of lower rated tranches has developed over time, are also a case in point.

As exposure aggregation on a borrower basis is a prerequisite not only for the HHI calculation but for the measurement of name concentration in general, these technical difficulties require careful attention. Conversely, if this aggregation problem is already the most difficult obstacle to overcome in practice, it may be preferable to use more refined measures than the HHI if the additional data collection costs are small compared with the aggregation problem and if the information gain is substantial.

Apart from its widespread use in other areas, the HHI has also a firm foundation in model theory. It turns out that in special cases a granularity adjustment for the IRB model is a linear function of the HHI. A granularity adjustment that incorporates name concentration in the IRB model was already included in the Second Consultative Paper of Basel II[8] and was later significantly refined by the work of Martin and Wilde[9] and further in Gordy and Lütkebohmert (2007). Given a portfolio of N borrowers, Gordy and Lütkebohmert developed the following simplified formula for an add-on to the capital for unexpected loss (UL capital)[10] in a single-factor model:

$$GA_N = \frac{1}{2K_N^*} \sum_{n=1}^{N} w_n^2 LGD_n \left[\delta \left(K_n + LGD_n PD_n \right) - K_n \right] \tag{5.1}$$

where w_n denotes the relative portfolio share of the exposure to the n-th borrower, K_n the UL capital for this exposure, LGD_n the expected loss-given-default, PD_n the PD, $K_N^* = \sum_{n=1}^{N} w_n K_n$ and δ is a constant parameter.[11] From Equation 5.1 follows immediately that the granularity adjustment is linear in the HHI if the portfolio is homogeneous in terms of PD and LGD.

The simplified Equation 5.1 follows from the 'full' granularity adjustment of Gordy and Lütkebohmert (2007) if quadratic terms are dropped. An alternative interpretation is to assume that any idiosyncratic risk in recovery rates that is still explicitly captured by the 'full' adjustment formula is eliminated by diversification.

An arguably severe obstacle to the application of earlier approaches that was resolved in the most recent proposal by Gordy and Lütkebohmert was the requirement to aggregate across all borrowers in a bank portfolio. The authors showed how the granularity adjustment can be conservatively approximated by considering only a part of the portfolio: namely, borrowers with exposures above a certain minimum exposure limit.

From a practitioner's perspective, the theoretical justification to compute the granularity adjustment from a subset of 'largest exposures' in the portfolio is potentially even more important than the quite parsimonious structure of Equation 5.1 compared with earlier proposals. More precisely, the required inputs for the approximation based on a subset of the portfolio are

1. The figures w_n, K_n, PD_n and LGD_n for a subset of $\overline{N} \leq N$ borrowers that have the highest capital contributions $w_n K_n$.

2. An upper bound of the exposure size for all $N - \overline{N}$ borrowers that are not included in this subset.
3. The total sum of UL capital K_N^* and expected loss EL_N^* for the credit portfolio.[12]

In various simulation exercises, based on empirically founded, realistic risk parameters, Gordy and Lütkebohmert find that their granularity adjustment is quite accurate even for portfolios of only 500 obligors, which is a relatively small number for a typical bank portfolio. Although its accuracy deteriorates for very small portfolios, the approximation is conservative, and for low-quality portfolios, it performs well for even smaller credit portfolios containing only 200 borrowers.

The convenient simplicity of the granularity adjustment formula comes at the price of a potential model error. The reason is that the granularity adjustment, unlike the IRB model, was developed in a CreditRisk$^+$ setting, and the CreditRisk$^+$ model[13] differs in the tail of the loss distribution from the IRB model. For this reason, their formula comes close to, but is not fully consistent, with the IRB model. It can be argued that this issue is more a lack of scientific elegance than a problem in practice, as the benefits from the tractability outweigh the costs of the model mismatch.

The authors briefly discuss other approaches to add a granularity adjustment to the IRB model but conclude that their approach is particularly suited for use in practice if the criteria of tractability and accuracy are considered together. Furthermore, the maturity adjustments of the IRB model can easily be accommodated by including them in the inputs.

4. Measuring sectoral concentration

Sectoral concentration can refer to an unbalanced distribution across geographical regions or business sectors. In standard portfolio models applied in the industry, this distinction is not particularly important, as country and industry are similarly captured by common risk factors. Unlike the development of a granularity adjustment, sectoral concentrations have until recently met with much less interest in the literature and in the industry. A notable exception is country risks, which refer to sectoral concentrations in geographical regions. In the following, the focus is on sectors, defined as branches of industry, but the methodology carries through to geographical sectors.

Unfortunately, no adjustment to the IRB model has been developed yet that accounts for sectoral concentrations and that was derived with the same rigor as the granularity adjustment of Gordy and Lütkebohmert (2006). As demonstrated by one of the methods discussed in this section, it is, however, possible to derive a multi-factor adjustment to a single-factor model at the price of extending this model to a multi-factor model and introducing additional correlation parameters.

Although sectors are directly linked to risk factors in the following, it is important to distinguish the 'sector effect' (more precisely 'sectoral concentration effect') on UL capital or VaR, which results from an unbalanced exposure distribution across sectors from a 'multi-factor effect' which derives from replacing the one-factor by a multi-factor model structure. The former effect results from changing the exposure distribution in the portfolio by increasing sector concentration. The latter effect is caused by an 'approximation' of the multi-factor model by a one-factor model without changing the portfolio. The multi-factor effect can be relatively small if the factor loadings in the single factor model are defined appropriately. The factor reduction technique of Andersen *et al.* (2003) can serve as an example. The effect of sector concentration instead can be substantial as was found in the empirical work by Duellmann and Masschelein (2006) and Heitfield *et al.* (2006).

A multi-factor asset-value model is used in the discussed methods as a general framework to account for sectoral concentrations. For simplicity, every borrower can be uniquely assigned to one sector that is represented by one common factor. In a multi-factor default-mode CreditMetrics-type model, the asset value return over the risk horizon or more general the default trigger Y_n is defined as

$$Y_n = \sqrt{r_{s(n)}} X_{s(n)} + \sqrt{1 - r_{s(n)}} \varepsilon_n \qquad (5.2)$$

where $s(n)$ denotes the sector to which the n-th borrower belongs, and $X_{s(n)}$ and ε_n are defined as the sector-specific common risk factor and the idiosyncratic risk factor, both independently standard normally distributed.

Correlations between the sector factors are collected in the correlation matrix Ω. A standard method of determining the sector correlations is to estimate them from stock index returns. In jurisdictions in which capital markets are less developed, this method becomes difficult to apply or the estimates may at least be less reliable for sectors with only a very few listed companies. Inferring default correlations from rating migrations or historical default rates can offer a solution but both methods are prone to their specific problems, such as the lack of external ratings in some jurisdictions or the scarcity of default events, for which unrealistically long time series are required. In the following, I will abstract from these empirical challenges and treat sector (factor) correlations as given. As the quantification of concentration risk is part of Pillar 2, it is fair to state that less strict standards need to be met than for a Pillar 1 model.

Even if sector correlations are available, the calculation of UL capital in the presence of sectoral concentrations encounters two problems. The first problem concerns the calculation procedure itself. This is usually done by Monte Carlo simulation but in practice tractable and robust approximations are desirable, particularly for banks that do not have a portfolio model with a simulation engine in place. The second problem occurs when comparing the output of the multi-factor model with the capital figure produced by the IRB model. The difference depends on the parameter settings in the IRB model, and it is not necessarily fully explainable by the sector allocation.

A few modelling approaches have been put forward in recent literature to account for sectoral concentrations. In the following, I focus on three of these models, which are used in Garcia Cespedes *et al.* (2006), Pykhtin (2004) and Duellmann and Masschelein (2006). All three approaches employ at some stage a multi-factor model, either through a calibration exercise or as a basis for a closed form approximation of UL capital. They were selected because they look appealing in terms of tractability and applicability in high dimensions.

A key contribution of Garcia Cespedes *et al.* (2006) is the development of a diversification factor DF that works as a scaling factor to the sector aggregate of UL capital $K_s^{*,sf}$

or sector s, in a single-factor model and produces an approximation of the UL capital $K_N^{*,GC}$ for the total portfolio.

$$K_N^{*,GC} = DF(CDI, \overline{\omega}) \sum_{s=1}^{S} K_s^{*,sf}. \tag{5.3}$$

The diversification factor $DF(.)$ is defined as a second-order polynomial $f(.)$ of the capital diversification index CDI and a weighted average sector correlation $\overline{\omega}$.[14] The surface parameterization of $f(.)$ is calibrated to a multi-factor model by minimizing the squared error between $K_N^{*,GC}$ from Equation 5.3 and a capital figure computed by Monte Carlo simulations for a wide range of portfolios. Note that the aggregation of $K_s^{*,sf}$ is carried out on a sector level, implicitly assuming a fine granular portfolio.

An alternative approach that offers an analytic approximation formula has been presented by Pykthin (2004) and was used in a simplified version in Duellmann and Masschelein (2006). Pykhtin's model is based on previous work by Gourieroux et al. (2000) and Martin and Wilde (2002). In terms of the applied methodology, it is closely linked to the granularity adjustment of Gordy and Lütkebohmert (2007) as it also employs a Taylor series approximation of the VaR.[15]

The simplifications by Duellmann and Masschelein from Pykhtin (2004) lie firstly in applying the approximation formulae to sector aggregates instead of at single-borrower level and secondly in dropping the second-order term of Pykhtin's VaR approximation. Under these assumptions, UL capital can be approximated by

$$K_N^{*,DM} = \sum_{s=1}^{S} w_s^* \overline{LGD}_s \left[\Phi \left(\frac{\Phi^{-1}(\overline{PD}_s) - \sqrt{\rho_s^*}\Phi^{-1}(\alpha_q(X))}{\sqrt{1-\rho_s^*}} \right) - \overline{PD}_s \right], \tag{5.4}$$

where w_s^* denotes the sum of single-exposure weights, expressed as relative shares of the portfolio exposure and $\alpha_q(X)$ the q-quantile of the distribution of X. $\overline{LGD}_s$ and $\overline{PD}_s$ are average values for sector s. Note that the systematic factor X is the same for all sectors but ρ_s^* is sector-dependent. The asset correlations ρ_s^* are calculated as proposed in Pykhtin (2004), broadly speaking by maximizing the correlation between the single systematic risk factor in the simplified model and the original sector factors.[16]

For their analysis of the approximation error, the authors employ realistic portfolios, constructed from German central credit register data, and an 11-sector (factor) model with empirically founded factor correlations. For fine-grained portfolios with homogeneous PDs and LGDs in every sector, the approximation error ends up moving in a close range of $\pm 1\%$. The approximation error is measured relative to UL capital obtained from Monte Carlo simulation-based estimates of UL capital. The effect of the absence of the second and higher order terms in Equation 5.4 is negligible. If dispersion in PDs is introduced in each sector, $K_N^{*,DM}$ provides a conservative estimate of UL capital because of the concave relationship between UL capital and PD.

Duellmann and Masschelein observe that the approximation error increases if they consider less granular portfolios that are heterogeneous in the size of single exposures and broadly representative for some small, regional German banks. Even in these cases, the absolute approximation error for UL capital stays well below 10%, measured relative to the simulation-based estimate, and Equation 5.4 still produces a conservative approximation.[17] The reason is that the upward bias from using sector-average PDs instead of borrower-dependent PDs exceeds the downward bias from neglecting the lower granularity.

A comparison of the approximation formulae used by Garcia Cespedes *et al.* (2006) Pykthin (2004) and Duellmann and Masschelein (2006) shows that all three have different strengths and weaknesses. The idea of Garcia Cespedes *et al.* to develop a scaling factor to UL capital $(K_s^{*,sf})$ that in turn is based on a single-factor model appears to be convenient: As the single-factor model has a simple, closed form solution for UL capital the computational burden is greatly reduced.

Caution is warranted in the parameterization of the model from which $K_s^{*,sf}$ is computed, as the scaling factor is restricted to the interval from zero to one and, therefore always produces a capital relief relative to $K_s^{*,sf}$. Consider, for example, a portfolio in which all exposures are concentrated in a single business sector $(CDI = 1)$, thus making the scaling factor one. $K_s^{*,sf}$ should be interpreted in this case as the UL capital required for a portfolio that is fully concentrated in the same business sector. As a consequence this UL capital figure should be based on intra-sector asset correlations. The asset correlations in the IRB model are instead best understood as an *average* of intra-sector and inter-sector asset correlations. As inter-sector correlations are typically lower than intra-sector correlations, the asset correlation of exposures in the same sector generally should be higher than in the IRB model. Therefore, if $K_s^{*,sf}$ were computed by the IRB model, diversification would be double-counted and the capital amount generally would be lower than adequate for exposures in the same sector.

In other words, to justify the capital relief from diversification, measured by $DF(.)$, it would be necessary to scale the asset correlations in the single-factor model up from those of the IRB model. As the intra-sector correlations that are consistent with the calibration of the IRB model are not available, it is not possible to achieve consistency with the IRB calibration. Therefore, it is also not recommendable to use the multi-factor adjustment of Garcia Cespedes *et al.* with $K_s^{*,sf}$ based on the IRB model to compute a diversification adjustment to the IRB model. Consistency with the single-factor model requires that $K_s^{*,sf}$ is properly parameterized, as will be demonstrated by an example in the next section.

The Pykthin model and the simplified version in Duellmann and Masschelein (2006) have the conceptual advantage of not requiring a calibration exercise as the model of Garcia Cespedes *et al.* (2006). Furthermore, if Pykhtin's original approximation formula is applied, granularity is automatically captured, i.e. if the aggregation in Equation 5.4 is carried out over borrowers with individual PDs instead of with sectoral averages. A further gain in accuracy is brought about by adding a higher order adjustment term. For large portfolios, the numerics can, however, become tedious. If the input parameters are instead computed at the coarser sector level, the methods of Duellmann and Masschelein and Garcia Cespedes *et al.* share the same restriction of not considering the granularity of the portfolio. The approximation accuracy therefore becomes a key criterion in the evaluation of $K_N^{*,GC}$ and $K_N^{*,DM}$ that will be examined in the next section.

Two other approaches that account explicitly for sectoral concentration have been developed in Tasche (2006) and Duellmann (2006). Tasche's approach is related to the work by Garcia Cespedes *et al.* (2006) and gives a more rigorous foundation of the diversification factor. Although he offers an analytic solution for the two-factor case, his model becomes less tractable in higher dimensions. The infection model developed in Duellmann (2006) is a generalization of the binomial expansion technique.[18] The model parameters are calibrated to a multi-factor CreditMetrics-type model, but simulation results suggest that the model is quite sensitive to the confidence level for which it was calibrated.

Contrary to the granularity adjustment of Gordy and Lütkebohmert (2007), the three models that account for sector concentrations are separate models, not direct extensions of the IRB model, and none of the authors discusses an extension into a mark-to-market valuation setting that would also require maturity adjustments.

5. Numerical example

The following example demonstrates how UL capital can be approximated by the three methods put forward in Garcia Cespedes et al. (2006), Pykthin (2004) and Duellmann and Masschelein (2006). For this purpose, the three approximation formulae are applied to credit portfolios with realistic risk characteristics. A simulation-based estimate for UL capital serves as benchmark.

The portfolio allocation to 11 business sectors is presented in Figure 5.1 and taken from Duellmann and Masschelein (2006). The inter-sector asset correlations were estimated from MSCI stock index returns for the respective sectors (average over sectors: 0.14) and the intra-sector asset correlation r_s is set to 0.25 for all sectors. The benchmark portfolio represents the sector distribution obtained from an aggregation of large exposure portfolios in the German banking system, based on the German central credit register. The concentrated portfolio is much more concentrated in a single sector (Capital goods) but still reflects a degree of sectoral concentration observed in real banks.[19]

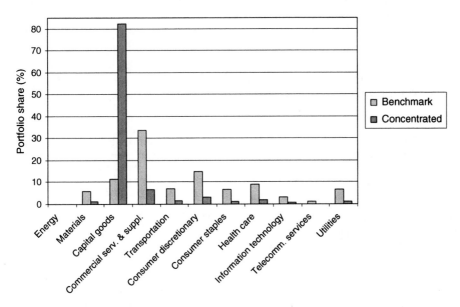

Figure 5.1 Exposure distribution across sectors of the benchmark portfolio (portfolio 1) and the more concentrated portfolios (portfolio 2 and 3)

There are four ways to calculate UL capital:

1. An estimate from $2 \cdot 10^6$ Monte Carlo simulation runs in the multi-factor model that serves as a benchmark for the following three methods.
2. The approximation formula from Pykthin (2004), which differs from Equation 5.4 in that the aggregation is carried out over single borrowers instead of sectors and includes a second-order multi-factor adjustment term.
3. Approximation formula (Equation 5.4) [Duellmann and Masschelein (DM) formula]
4. Approximation formula (Equation 5.3) [Garcia Cespedes (GC) *et al.* formula].

Note that for the approximation formula of Garcia Cespedes *et al.* (2006), $K_s^{*,sf}$ is not computed with the asset correlations from the IRB model. Instead, the parameter ρ in Equation 5.10 in Appendix A.1 is replaced by the intra-sector correlation r_s, which is the same in all sectors. Its value of 0.25 always exceeds the asset correlations in the IRB model, which vary between 0.12 and 0.24, depending on the PD.

Only portfolios with homogeneous PD and LGD are considered in the example, as the approximation formulae by Garcia Cespedes *et al.* (2006) and Duellmann and Masschelein (2006) do not account for heterogeneity in these input variables. The LGD is set to 45% and the PD to 2% for all exposures. Differences in the granularity of the exposures are also not accounted for by these two models, yet this effect is included in the analysis because low granularity creates a concentration risk.

Table 5.1 summarizes UL capital figures for three portfolios. The entries in the column for portfolio 1 refer to a portfolio that is distributed across sectors as the benchmark portfolio (see Figure 5.1) and that comprises 600 borrowers with an exposure size of 1/600 to each borrower. The next column contains the results for portfolio 2 that has the sectoral distribution of the concentrated portfolio in Figure 5.1 but is still as finely granular as portfolio 1. The last column collects the results for portfolio 3, which has the same concentrated sectoral distribution as portfolio 2 but a higher name concentration than the other two. It was generated from the second portfolio by replacing 384 of the 493 exposures of 1/600 in the Capital goods sector by 32 lumpy exposures of 1/50. The lump size of 1/50 is motivated by European large exposure rules[20] as this number is the maximum exposure limit of 25% of regulatory capital, assuming a regulatory capital

Table 5.1 Comparing UL capital in a multi-factor model with analytic approximations

	Portfolio		
	1	2	3
Exposure distribution across sectors:	Benchmark	Concentrated	
Number of borrowers (exposures)	600	600	248
UL capital & (difference from MC estimate)[a]			
MC simulation	7.88	10.8	11.63
Pykthin formula	7.97 (1.1)	10.81 (0.1)	11.70 (0.6)
DM formula	7.75 (−1.7)	10.67 (−1.2)	10.67 (−8.3)
GC *et al.* formula	8.08 (2.5)	10.74 (−0.6)	10.74 (−7.7)

[a]UL capital in percent of total exposure; difference in brackets as a percentage of MC estimate.

charge of 8% of the portfolio's nominal value. The number of 32 lumpy exposures with a total of 32/50 still satisfies the maximum limit of eight times the regulatory capital set by the same rules. The coarser granularity of the second portfolio is reflected in the higher HHI value of 0.0134, compared with a value of 0.0017 for the first.

If we compare the UL capital estimates in Table 5.1 for portfolio 1, i.e. the benchmark portfolio, this reveals a relatively high accuracy for all three approximation formulae, as the errors measured relative to the simulation-based estimates lie in a narrow band between -1.7% and 2.5%. Given that the GC *et al.* formula relies on a calibration on a set of different portfolios with other risk characteristics, the numerical example is, in fact, an out-of-sample test with encouraging results.

A similar degree of accuracy also holds for all three approximations in the case of the second, sectorally more concentrated portfolio. The increase in UL capital from the benchmark portfolio to the concentrated portfolio amounts to 37.1% for the simulation-based estimates and confirms the need to consider sectoral concentrations in credit portfolios.

The last column collects the UL capital estimates for the less granular portfolio 3 of only 248 borrowers. The simulation-based value increases by 7.7% relative to the second, more granular portfolio that has the same sectoral distribution. The DM and GC *et al.* formulae do not account for granularity, therefore causing the error to increase in this range. The Pykthin formula instead accounts for the lower granularity. Its relative error of 0.6% is similar to the errors for the highly granular portfolios and much lower than for the other two approximation formulae.

In summary, the numerical example shows that the two approximation formulae by Garcia Cespedes *et al.* (2006) and Duellmann and Masschelein (2006) both perform very well if the portfolio is homogeneous in terms of PD and LGD and if the portfolio is highly fine-grained. Both fail to capture name concentration, in which case the Pykthin (2004) model still offers a viable alternative to running Monte Carlo simulations. Its technical burden could be a main pitfall of Pykthin's approximation formula if it is applied on a single-borrower level: In the above example, its computation took nearly as long as the Monte Carlo simulations, and computation time for this method increases faster than for the Monte Carlo simulations with the same number of borrowers.

6. Future challenges of concentration risk measurement

The various measurement approaches discussed in the previous three sections have the same methodological underpinning as typical credit risk models currently applied in the industry and therefore are subject to the same limitations. Recent empirical research suggests the existence of credit risk dependencies, which cannot be explained by typical multi-factor models which assume independence conditional on the common risk factors.[21] One economic explanation of these dependencies is the existence of business links between firms, which may become particularly relevant for exposure concentrations in large firms. As no model standard has been established yet that captures such contagion effects, incorporating them remains a challenge.

UL capital is quite sensitive to the confidence level of the VaR calculation. Figure 5.2 shows simulation-based estimates of UL capital for confidence levels between 95% and

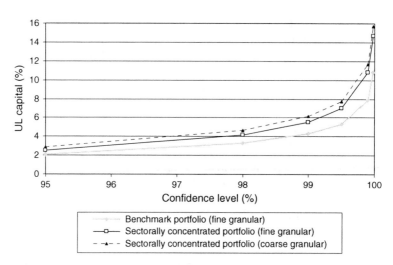

Figure 5.2 UL capital for the fine-grained benchmark portfolio and the fine-grained and coarse granular, (sectorally) concentrated portfolio at different confidence

99.98% for two portfolios: the benchmark portfolio and the concentrated portfolio from the numerical example in the previous section.

UL capital increases strongly with the confidence level, and this increase is notably stronger for the more concentrated portfolios. The difference in UL capital between the two fine-grained portfolios, measured relative to UL capital for the benchmark portfolio, increases from 18% for a 95% confidence level to 36% for the 99.98% confidence level. The difference in UL capital between the fine-grained and the coarse granular sectorally concentrated portfolios also increases with the confidence level, but contributes much less to the cumulative effect of sectoral and name concentration. The reason for the stronger increase of UL for the two more concentrated portfolios is that higher confidence levels implicitly give more weight to concentration effects. This example demonstrates that management decisions such as setting the confidence level and quantifying the capital impact of risk concentrations are intricately linked.

Not only management variables but also estimates from external data can have a strong impact on measuring the impact of risk concentrations. The presented methods suffer from the widely known weaknesses of most existing industry models, such as correlation estimates being either unstable over time or inaccurate, the latter due either to data constraints or a lack of recognition of dependencies between PD and LGD. A potential remedy is offered by stress tests for risk concentrations.

As concentration risk is implicitly captured by standard industry models, it is often difficult to distinguish between stress tests for concentration risk and (more general) stress tests for credit risk. A forecast of a sudden drop in the production index of the car industry can serve equally well as an example of a concentration risk stress test – if the portfolio is known to be concentrated in the automobile sector – as well as of a more general credit risk stress test. In Pillar 2, banks are encouraged to conduct periodic stress tests of their major credit risk concentrations.[22]

The stress test procedure in Bonti et al. (2006) offers a potential means of uncovering sectoral concentrations and assessing the resulting portfolio risk. Recognizing that the exact stress event cannot be forecasted, the authors restrict the range of a specific risk factor such that the expectation value of the stressed factor in the truncated subspace corresponds to the 'expected' stress scenario. Staying within their model framework and employing the original correlation matrix, their stress test methodology also picks up losses under stress conditions that are caused by correlations between sectors.

A technical burden that their stress test methodology shares with various other approaches is that each stress test requires new Monte Carlo simulations. The UL capital approximation in Garcia Cespedes *et al.* (2006) offers a framework that can also be used for stress testing and that greatly reduces this technical burden.[23]

Finally, given the progress that has been made in credit risk modelling, data constraints have become one of the most important challenges for credit risk management in practice. They can be relatively basic requirements such as being able to aggregate all exposures to the same borrower on a group level, or they can be more sublime issues such as the stability of correlation estimates and how to capture observed dependencies that are no longer explained by standard models based on the conditional independence assumption.

7. Summary

Concentration risk in credit portfolios is woven into the credit risk of the portfolio and, therefore, implicitly accounted for in best-practice multi-factor portfolio models applied in the industry. This paper gives an overview and compares various approximation methods that measure the capital impact of risk concentrations in the context of the single-factor model that underlies the IRB risk-weight functions. For this purpose, risk concentrations are understood as violations of two specific assumptions of the IRB model: firstly, that the portfolio is highly granular in terms of exposures to individual borrowers and, secondly, that systematic credit risk is fully captured by a single risk factor. Measuring the capital impact of any violation of the second assumption must also take into account the fact that the IRB model was calibrated to produce a UL capital charge that would be appropriate for large credit portfolios that are well-diversified across sectors (risk factors).[24] As a consequence, the regulatory minimum capital requirements can underestimate the prudential amount of capital needed for credit portfolios that have either high name concentration or an unbalanced sectoral exposure distribution, either across geographical or across business sectors.

Recent empirical work by Duellmann and Masschelein (2006) and others confirms a widely held perception that both types of risk concentration can have a material impact on total portfolio risk and that sectoral concentrations typically have a stronger effect. Neither name nor sectoral concentrations can be accounted for by model adjustments that are fully consistent with the IRB model framework. In other words, none of the discussed methods promise to directly adjust the IRB model in a way that is fully consistent with this model. As concentration risk is treated under Pillar 2 of the new Framework, it can be argued that full consistency is not needed.

In the case of name concentration, the granularity adjustment developed in Gordy and Lütkebohmert (2007) comes very close to achieving consistency with the IRB model. The option of determining a simplified granularity adjustment that requires the exposure aggregation per borrower only for a portfolio subset of the largest borrowers significantly facilitates its application in practice.

Sectoral concentrations – or, equivalently, violations of the single-factor assumption – are more difficult to address than name concentration. A multi-factor framework in which the impact of the portfolio distribution across sectors is captured by systematic risk factors can account for concentration risk in geographical or business sectors. It forms also the basis of the approximation formulae developed in Garcia Cespedes et al. (2006) and Duellmann and Masschelein (2006). Their methods greatly reduce the technical burden of calculating a UL capital figure, which accounts for the exposure distribution across sectors. The approach by Garcia Cespedes et al. (2006) captures risk concentrations by a scaling factor to a capital figure, which in turn is based on a single-factor model. This single-factor model differs from the IRB model in its parameterization because it assume that all exposures are located in the same business sector. This assumption implies that asset correlations in this model generally should be higher than in the IRB model

In summary, neither methodology can provide a fully consistent adjustment to the IRB model, as the intra-sector and inter-sector asset correlations used in its calibration are not available. In many cases, this may not be necessary as the discussed methods can at least provide a ranking of credit portfolios in terms of economic capital for sector concentration, which already gives important information.

Notwithstanding these limitations of quantifying concentration risk in credit portfolios by simple, fit-for-purpose tools, a model-based approach like the one followed in the methods presented in this paper can offer substantial benefits over ad hoc measures in terms of accuracy and robustness of the results. Considering that the exposure aggregation at borrower level already poses a major challenge in practice and that this hurdle also concerns simple ad hoc measures such as the HHI, the cost of additional data requirements for model-based methods may appear justified considering the achieved information gain.

Identifying risk concentrations and quantifying their impact on portfolio risk is only the first step in concentration risk management. The second, and equally important, step – which is, however, beyond the scope of this paper – concerns the prudential management of risk concentrations. The importance of the management aspect is also emphasized in Pillar 2: Banks should review the results of their stress tests on concentration risk and respond to potential changes in market conditions that could adversely impact the bank's performance. Likewise, they should ensure that, in respect of credit risk concentrations, risk management complies with the Basel Committee on Banking Supervision (BCBS) document 'Principles for the Management of Credit Risk'.[25]

References

Andersen, L., Sidenius, J. and Basu, S. (2003). All your hedges in one basket. *Risk Magazine*, **16** (November) 67–72.

Basel Committee on Banking Supervision (2000) Principles for the Management of Credit Risk. Basel, September (http://www.bis.org/publ/bcbs75.pdf).

Basel Committee on Banking Supervision (2001) The New Basel Capital Accord, Consultative document, Basel, January.

Basel Committee on Banking Supervision (2004) International Convergence of Capital Measurement and Capital Standards: A Revised framework (http://www.bis.org/publ/bcbs107b.pdf).

Basel Committee on Banking Supervision (2005) Update on Work of the Accord Implementation Group Related to Validation under the Basel II Framework. Newsletter no 4, January (http://www.bis.org/publ/bcbs_nl4.htm).

Basel Committee on Banking Supervision (2006) Studies on Credit Risk Concentration (http://www.bis.org/publ/bcbs_wp15.pdf).

Bonti, G., Kalkbrener, M., Lotz, C. and Stahl, G. (2006). Credit risk concentrations under stress. *Journal of Credit Risk*, **2**(3), 115–36.

Cifuentes, A. and O'Connor, G. (1996). The binomial expansion method applied to CBO/CLO analysis. In *Moody's Investor Service*, Special Report, December.

Collin-Dufresne, P., Goldstein, R.S. and Helwege, J. (2003). *Is Credit Event Risk Priced?* Modeling contagion via the updating of beliefs, Carnegie Mellon University Working Paper.

Credit Suisse Financial Products (1997) CreditRisk⁺: A Credit Risk Management Framework, Technical document (http://www.csfb.com/creditrisk).

Das, S.R., Duffie, D., Kapadia, N., *et al.* (2006). Common failings: How corporate defaults are correlated. NBER Working Paper No. W77967.

Duellmann, K. (2006). Measuring business sector concentration by an infection model. Deutsche Bundesbank Discussion Paper (series 2), no 3.

Duellmann, K. and Masschelein, N. (2006). Sectoral concentration risk in loan portfolios and economic capital. Deutsche Bundesbank Discussion Paper (series 2), no 9 and National Bank of Belgium Working Paper, no 105.

Garcia Cespedes, J.C., de Juan Herrero, J.A., Kreinin, A., *et al.* (2006). A simple multi-factor adjustment, for the treatment of diversification in credit capital rules. *Journal of Credit Risk*, **2**, 57–85.

Gordy, M. (2003). A risk-factor model foundation for ratings-based bank capital rules. *Journal of Financial Intermediation*, **12**, 199–232.

Gordy, M. and Lütkebohmert, E. (2007). Granularity adjustment for Basel II. Deutsche Bundesbank Discussion Paper (series 2), no 1.

Gourieroux, C., Laurent, J.-P. and Scaillet, O. (2000). Sensitivity analysis of values at risk. *Journal of Empirical Finance*, **7**, 225–45.

Gupton, G., Finger, C. and Bhatia, M. (1997). CreditMetrics - Technical Document.

Heitfield, E., Burton, S. and Chomsisengphet, S. (2006). Systematic and idiosyncratic risk in syndicated loan portfolios. *Journal of Credit Risk*, **2**, 3–31.

Loesch, S. (2006). Concentrating on premium. Managing credit portfolio inefficiencies, JPMorgan EMEA Structured Finance Advisory, July.

Martin, R. and Wilde, T. (2002). Unsystematic credit risk. *Risk Magazine*, **15** (November), 123–8.

Pykhtin, M. (2004). Multi-factor adjustment. *Risk Magazine*, **17** (March), 85–90.

Tasche, D. (2006). Measuring diversification in an asymptotic multifactor framework. *Journal of Credit Risk*, **2**, 33–55.

Tornquist, P. (2004). Concentration Risks Remain High At European Banks, Standard & Poor's (http://www.ratingsdirect.com).

Wilde, T. (2001) Probing granularity. *Risk Magazine*, **14** (August), 103–6.

Notes

1. See BCBS (2005), p. 2.
2. See BCBS (2004), para. 773.
3. See BCBS (2004), para. 771.
4. See BCBS (2004), para. 771.
5. A more technical presentation of the two assumptions and their relationship to name concentration and sectoral concentration is given in Appendix A.1.
6. See Tornquist (2004).
7. See Loesch (2006), p. 4.
8. See BCBS (2001).
9. See Wilde (2001) and Martin and Wilde (2003).
10. UL capital is defined as the difference between the value-at-risk and the expected loss. See Appendix A.1 for more details on UL capital in the context of a single-factor model.
11. The constant δ depends on the VaR confidence level and also on the variance of the systematic factor. Gordy and Lütkebohmert suggest a value of 5 as a meaningful and parsimonious number. An asterisk signals that the variable is computed as the sum of other variables.

12. For a mathematical definition of EL_N^*, see (5.9) in Appendix A.1.
13. See Credit Suisse Financial Products (1997).
14. See Appendix A.2 for further details on the definition of DF(.), its two components CDI and $\bar{\omega}$ and the DF surface parameterization.
15. For the relatively complex approximation formula, see Pykthin (2004).
16. See Appendix A.3 for a formal definition of ρ_s^*.
17. This is not contradicted by Table 5.1, in which the UL capital from their formula is slightly less than the simulation-based estimate, because the conservativeness is only due to PD heterogeneity, which is not present in that table.
18. See Cifuentes and O'Connor (1996).
19. For further details about the portfolio and the correlation estimates, see Duellmann and Masschelein (2006).
20. See Directive 93/6/EEC of 15 March 1993 on the capital adequacy of investment firms and credit institutions.
21. See, for example, Collin-Dufresne et al. (2003) or Das et al. (2007).
22. See BCBS (2004), para. 775.
23. See Garcia Cespedes et al. (2006) for further details.
24. See BCBS (2006).
25. See BCBS (2004), para. 775–6.
26. See Gupton et al. (1997).
27. This structural resemblance is misleading in that the calibration of the maturity adjustments, which is omitted here, was actually carried out in a mark-to-market framework.

Appendix A.1: IRB risk weight functions and concentration risk

Let q be the confidence level, $\alpha_q(Z)$ the q-quantile of the distribution of a random variable Z, N the number of borrowers and L_N the random loss rate of the total portfolio over a fixed risk horizon. In the limit, portfolio invariance requires that

$$\alpha_q(L_N) - \alpha_q(E[L_N|X]) \xrightarrow{N \to \infty} 0, \tag{5.5}$$

where $E[L_N|X]$ is defined by aggregating the expected loss conditional on the systematic risk factor X over all borrowers:

$$E[L_N|X] = \sum_{i=1}^{N} w_n LGD_n p_n(X), \tag{5.6}$$

where $p_n(X)$ denotes the PD conditional on X. Convergence in Equation 5.5 can also be achieved in the multi-factor case if X denotes a vector of common risk factors. The additional relationship

$$P(L_N \le E[L_N|\alpha_q(X)]) \xrightarrow{N \to \infty} q, \tag{5.7}$$

which leads to a simple formula for $E[L_N|X]$, however, only exists for a single-factor model. This explains why there is no simple 'multi-factor adjustment' that would allow us to account for an unbalanced distribution across sectors while retaining an analytic solution for UL capital.

In finite portfolios, the difference of $\alpha_q(L_N) - \alpha_q(E[L_N|X])$ can be interpreted as the impact of undiversified idiosyncratic risk. It describes the 'true' granularity adjustment that does not have an analytical solution. However, it can be approximated by, for instance, the granularity adjustment in Gordy and Lütkebohmert (2006), which is derived from a Taylor series approximation, disregarding terms of higher than second-order.

The UL capital charge K_N^* for the total portfolio in the IRB model captures the *unexpected loss* and is defined by

$$K_N^* = \alpha_q(L_N) - EL_N^* \tag{5.8}$$

as the difference between the VaR $\alpha_q(L_N)$ and the expected loss

$$EL_N^* = E[L_N] = \sum_{n=1}^N w_n LGD_n PD_n \tag{5.9}$$

where w_n denotes the relative portfolio share of the exposure to the n-th borrower, LGD_n the expected LGD and PD_n the borrower's PD. The structure of the IRB model is closely related to a one-period default-mode CreditMetrics[26] model.[27] Default is triggered if the ability-to-pay process Y_n crosses a default barrier c_n and Y_n can be decomposed into an idiosyncratic risk factor X and a systematic risk factor ε_n as follows:

$$Y_n = \sqrt{\rho}\, X + \sqrt{1-\rho}\, \varepsilon_n$$

where X and ε_n are independently standard normally distributed random variables. The parameter ρ, commonly referred to as *asset correlation*, is the correlation between the ability-to-pay processes of any pair of borrowers. The default barrier c_i is related to the (unconditional) PD by $PD_n = P(Y_n \le c_n) = \Phi(c_n)$, where $\Phi(.)$ denotes the cumulative standard normal distribution function.

In this model and by invoking Equation 5.7, *UL capital* K_n for the n-th borrower, given a risk horizon of 1 year and a confidence level q, can be calculated as follows:

$$K_n = E[l_n | \alpha_q(X)] = w_n LGD_n \left[\Phi\left(\frac{c_n - \sqrt{\rho}\, \Phi^{-1}(\alpha_q(X))}{\sqrt{1-\rho}} \right) - PD_n \right] \tag{5.10}$$

where l_n denotes the loss of the total exposure to the n-th borrower. The IRB risk weight formulae are based on Equation 5.10 multiplied by a constant of 12.5 to account for the minimum solvency ratio of 0.08. For simplicity, maturity adjustments, the dependency of ρ on PD in certain asset classes and the requirement to differentiate between an (expected) LGD and an economic downturn LGD are omitted in Equation 5.10.

Appendix A.2: Factor surface for the diversification factor

Garcia Cespedes *et al.* (2006) have calibrated the following factor surface $f(CDI, \overline{\omega})$ for the diversification factor.

$$f(CDI, \overline{\omega}) = 1 + a_{11}(1-\overline{\omega})(1-CDI) + a_{21}(1-\overline{\omega})^2(1-CDI) + a_{22}(1-\overline{\omega})^2(1-CDI)^2$$

with $a_{11} = -0.852$, $a_{21} = 0.426$ and $a_{22} = -0.481$,

$$CDI = \frac{\sum_{s=1}^S \left(K_s^{*,sf}\right)^2}{\left(\sum_{s=1}^S K_s^{*,sf}\right)^2} \tag{5.11}$$

and

$$\overline{\omega} = \frac{\sum\limits_{s=1}^{S} \sum\limits_{t=1,t\neq s}^{S} w_s w_t \omega_{st}}{\sum\limits_{s=1}^{S} \sum\limits_{t=1,t\neq s}^{S} w_s w_t} \tag{5.12}$$

with ω_{st} denoting the correlation between the s-th and t-th sector factor.

In Garcia Cespedes et $al.$ (2006), the correlation matrix of sector factors has only nine parameters instead of $S \cdot (S-1)/2$ in the general multi-factor model. The reason is that they define each sector factor as a weighted sum of a common risk factor for all sectors and a sector-specific risk factor. The correlation matrix is, therefore, fully specified by the S weights of the single common factor. In the numerical example, the correlation parameter ω_{st} in Equation 5.12 is taken instead from the more general matrix Ω.

Appendix A.3

In Pykhtin (2004), the ρ_s^* for $s \in \{1, \dots, S\}$ are determined as follows.

$$\rho_s^* = \sqrt{r_s} \sum_{t=1}^{S} \gamma_{st} b_t$$

where γ_{st} comes from a Cholesky decomposition of the factor correlation matrix Ω.

The coefficients $b_1, \dots, b_s$ and λ are given as the solution of the following $S+1$ equations:

$$b_t = \sum_{s=1}^{S} \frac{\theta_s}{\lambda} \gamma_{st} \text{ for } t \in \{1, \dots, S\}$$

$$\sum_{s=1}^{S} b_s^2 = 1$$

with

$$\theta_s = w_s^* \, \overline{LGD} \, \Phi \left(\frac{\Phi^{-1}\left(\overline{PD}_s\right) + \sqrt{r_s}\Phi^{-1}\left(\alpha_q\left(X\right)\right)}{\sqrt{r_s - 1}} \right).$$

6 A Simple method for regulators to cross-check operational risk loss models for banks

Wayne Holland* and ManMohan S. Sodhi*

Abstract

Under Basel II, banks are encouraged to develop their own advanced measurement approaches (AMA), allowing different approaches to flourish. However, this flexibility may pose a problem for the regulators in approving a particular bank's proposed capital requirement as indicated by the bank's own AMA model. We propose a simple method for regulators to cross-check any bank's proposed capital reserve using a lower bound function on the capital requirement. This bound is a function of the weight of the bank's estimate of its expected losses relative to the pooled loss experience of other banks. The regulator can use the bound to indicate potential problems with an individual bank's AMA model at the line of business level. We illustrate our approach with a bound that is easy to compute. Banks can also use this approach to cross-check their own models.

1. Introduction

We seek to provide a simple way for regulatory bodies to cross-check the operational risk loss model and the proposed capital requirement at the line of business level for a particular bank by using the pooled loss experience of other banks. For regulators, as well as for any bank developing its operational risk models using advanced measurement approaches (AMA), there are at least two challenges: (1) the variety of sophisticated statistical and other probability-based ideas for modelling and (2) the paucity of loss data available for constructing loss distributions. Our contribution is to provide regulators with a simple approach using a bank's expected losses as well as pooled loss data of other banks to cross-check this bank's method of estimating the capital reserve by line of business and in total.

Our motivation is to provide an approach that is somewhat between the Standard-ized Approach and the AMA in that we also seek minimum capital requirements at the line of business level. The Basic Approach sets minimum capital requirement as a per-centage, usually 15%, of annual gross income for the business as a whole for most banks. The Standardized Approach takes this further by providing percentages at the line of business level for most banks. The AMA is specific to an individual bank that uses their

* Cass Business School, City University London, UK

own approach to determine capital requirements for their different lines of business and for the bank as a whole. By contrast, our approach is standardized in the sense that it is transparent and uses the loss experience of all reporting banks. However, the regulators can apply our approach to a particular bank, knowing only the bank's annual gross income and expected losses at the line of business level.

In our approach, besides the proposed capital reserve, the regulator asks the bank to provide its expected annual loss and expected annual gross income by line of business and for the bank as a whole. Using a weight α that ranges from 1 down to 0 to combine the bank's loss parameters with those of the pool, the regulator obtains a lower bound function based on a particular standard probability distribution. (We use a lognormal distribution for illustration, but we can use other distributions as well.) The regulator uses the 99.9-th percentile of this distribution to obtain a lower bound for capital requirement for all values of α for each of the lines of business and for the bank as a whole.

If the bank's proposed capital requirement for any line of business or for the total is below the respective bound for all values of α, the regulators will have flagged a situation where they need to take a closer look at the bank's AMA. Even where all the lines of business have the bank's capital requirements by line of business exceed the respective lower bound for some values of α, very different ranges of α for different lines of business also indicate a possible problem.

Our approach is simple to use as a diagnostic tool. It uses not only the bank's estimates (subjective and/or objective) but also the pool of banks' operational losses collected by regulators, e.g. the 2002 losses reported by 89 banks, which the Basel Committee (2003) has summarized. Moreover, our method requires only the expected annual losses and annual gross income by line of business and in total, while using other parameters from the more reliable pooled loss experience of all reporting banks. Finally, our method uses standard probability distributions well accepted in operational risk modelling – doing so does not make their use 'correct' in any objective sense, but it does make it easier to explain their use.

On the contrary, our approach has many limitations stemming from the simplifying assumptions. (Therefore, regulators or banks should use our approach only to indicate potential problems with a bank's operational risk modelling rather than use our approach as an AMA *per se*.) We assume that the number of losses is Poisson-distributed with lognormally distributed impact per loss event. This is not a serious problem because many AMA models appear to make the same assumption in conjunction with subjective score-based methods. Furthermore, we assume that the pooled history of losses of reporting banks (as a percentage of revenues) is the same as having a longer history of a single bank. Although this makes our method less vulnerable to the vagaries of a particular bank's short history, we are assuming that all banks have the same coefficient of variation (standard deviation adjusted by the mean) of future annual losses scaled by annual gross income. We ignore the fact that the data compiled by the regulator do not include all losses, only those above thresholds like 10 000 euros (Basel Committee, 2003). We can adjust for that – the adjustment may be smaller than 5% – but we have not shown the adjustment. Finally, we use the lognormal distribution based on simulation rather than on theoretically derived results.

In Section 2, we discuss the Basel II requirements and approved approaches for calculating capital requirement. In Section 3, we present our approach. Sections 4 and 5 outline a justification of our method and present the simulation on which we relied to propose the lower bound based on the lognormal distribution. Section 6 concludes this paper and outlines areas of further work.

2. Background

Basel II capital requirements allow for three approaches for setting minimum capital requirement to cover operational risk losses (McNeil *et al.*, 2005: pp. 463–466). Of these, AMA are based on the development of probabilistic models to obtain the capital requirement.

AMA models range in sophistication and approach. Some models use extreme value theory (e.g. Cruz, 2003: pp. 63–85 or Embrechts *et al.*, 1997: pp. 283–358). Others fit probability distributions to describe the number of loss events per year and the severity of losses (e.g., Alexander, 2003: pp. 129–211) in the 8×7 different types of losses (Basel Committee, 2003). Bayesian networks have also generated some interest owing to the paucity of loss data and to a desire to have models built on causes rather than effects or reported losses only (Cowell *et al.*, forthcoming).

The supervisory framework introduced by the Basel Committee as Basel II rests on three pillars: minimum capital requirement, supervisory review and market discipline. For determining the minimum capital requirement, the first of these pillars, the Basel Committee allows three possible approaches (McNeil *et al.*, 2005: pp. 463–466):

1. *Basic Indicator Approach*: A bank can choose an indicator of operational risk such as gross income or revenues. The bank can then take a fixed percentage of this measure to be the Minimum Capital Requirement. The value of this percentage is typically 15% of annual gross income (averaged over 3 years), but in some cases a value as high as 20% may be more appropriate.
2. *Standardized Approach*: This is the same as the Basic Approach except that the bank applies potentially different factors to annual gross income for each of eight BIS-recognized lines of business. These lines of business are (1) corporate finance; (2) trading and sales; (3) retail banking; (4) commercial banking; (5) payment and settlement; (6) agency services and custody; (7) asset management and (8) retail brokerage. The respective values of the multiples for determining the Minimum Capital Requirement for these lines of business depend of the relative risk for each line.
3. *Advanced Measurement Approaches*: A bank may use more sophisticated probabilistic or other approaches based on internal measurements, subjective measures regarding controls in place and other bank's pooled loss experience. The framework requires banks not only to analyse its data by the eight business lines outlined under the Standardized Approach but also by seven standard loss event categories – (1) internal fraud; (2) external fraud; (3) employment practices and workplace safety; (4) clients, products and business practices; (5) damage to physical assets; (6) business disruptions and systems failures; and (7) execution, delivery and process management. The Basel Committee has not imposed restrictions on the type of approaches whether based on collection and modelling of internal data, scenario analysis or expert opinion.

Our method extends the Standardized Approach to cross-check the capital reserve required by any particular AMA. Regarding AMA, the Basel Committee has not been overly prescriptive in its framework. This allows room for new approaches to develop but places a burden on the regulators to verify a bank's internal approaches (Alexander, 2003: pp. 56–57). Moreover, these models require detailed loss data histories and herein lies the rub. Either organizations have not collected all loss data in all categories, or have collected loss data only above certain thresholds (say, only when the loss was above 5000 euros), or in categories that are different from the 56 that the Basel Committee has recommended for standardization. In this paper, we propose a validation procedure to assess the reasonableness of the outcome of the particular AMA that a bank may be using, taking advantage of the data of other banks available to the regulators.

3. Cross-checking procedure

In the following discussion, Greek letters denote the parameters of the appropriate distribution. To keep our notation simple, we do not use subscripts to distinguish the specific estimated parameters for a particular line of business or for the bank as whole. Nor do we distinguish between estimates obtained from the pooled data or in any other way and the (unknown) value of the parameters themselves using the traditional 'hat' symbol. This section describes the procedure whereas the following two sections (4 and 5) outline the justification.

Using the pool-of-loss experience reported by different banks, the regulator can compute μ_p and σ_p, the mean and standard deviation respectively of annual loss as a percentage of annual gross income. These two estimated parameters are for each of the eight lines of business as well as for a bank as a whole; therefore, there are nine pairs of parameters.

Using these parameters, the regulator can compute a lower bound function for the capital reserve for the bank in question for each line of business and for the bank as a whole as follows:

1. *Input from bank*: The regulator asks the bank to provide the expected annual loss and expected revenues for each line of business and for the total. This would give the regulator μ_b, the mean annual loss as a percentage of annual gross income (for line of business or bank) for bank b's line of business or grand total, respectively. (There are thus up to nine such parameters, less when a bank does not have a particular line of business from the eight listed by the Basel Committee.) Most banks will likely have some subjective measures based on managers' expertise to provide estimates of the expected loss (Table 6.1).
2. *Weighted mean and standard deviation*: Take α to be a parameter representing the relative weight to be given to bank b data in relation to the pooled data $(0 \le \alpha \le 1)$. Then, for each line of business and total, the regulator can define weighted parameters specific to the bank in so far as the value of α is close to 1.

$$\mu_\alpha = \alpha\mu_b + (1 - \alpha)\mu_p$$

$$\sigma_\alpha = \left(\frac{\sigma_p}{\mu_p}\right)\mu_\alpha \tag{6.1}$$

Table 6.1 Inputs that the regulator requires, in our approach, from the bank in question

Line of business	Expected annual loss	Annual gross income	Proposed capital requirement
Corporate finance			
Trading and sales			
Retail banking			
Commercial banking			
Payment and settlement			
Agency services and custody			
Asset management			
Retail brokerage			
BANK as a whole			

When α is close to 0, the two parameters approach the pooled values. As before, there are (up to) nine pairs of means and standard deviations.

3. *Parameters of bounding distribution*: For each line of business and for the bank as a whole, the regulator can then define parameters for the distribution used for bounding; in our case, the lognormal distribution for illustration. (We observe in the next section that a log-lognormal could provide a better bound.) For the lognormal distribution, if m and s are the mean and standard deviation of the underlying normal distribution, then

$$
m_\alpha = \log_e \left(\frac{\mu_\alpha^2}{\sqrt{\sigma_\alpha^2 + \mu_\alpha^2}} \right)
$$

$$
s_\alpha = \sqrt{ \log_e \left(\frac{\sigma_\alpha^2 + \mu_\alpha^2}{\mu_\alpha^2} \right) }
$$

(6.2)

4. *Lower bound function*: For each line of business and for the total, the regulator defines L_α to be a lower bound on the minimum capital requirement based on the 99.9-th percentile of the lognormal distribution less the expected value μ_α, respectively, as functions of α ranging from 0 to 1. In other words, L_α is such that $P(X < L_\alpha) = 0.999$ where $X \sim$ lognormal (m_α, s_α) (Figure 6.1).

5. *Decision*: If the bank's proposed capital requirement for any line of business or for the total is *below* our lower bound function for all values of α, the implication is *either* that the bank's AMA is providing low values of capital reserve *or* that the bank is significantly better than other banks in controlling operational risk losses. Either way, the regulator needs to examine the bank's AMA and loss experience more closely. Otherwise, the regulator could consider the bank's proposed capital reserve and AMA as having been cross-checked although the regulator may still want to understand the bank's approach better.

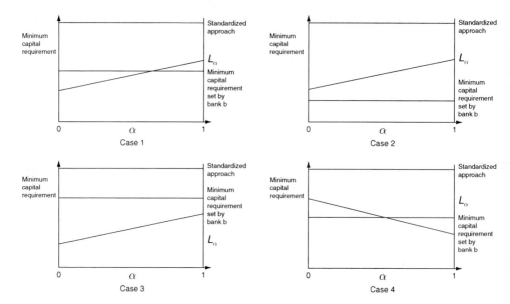

Figure 6.1 Four schematic cases to illustrate the decision procedure. In Cases 1 and 4, the bank's proposed capital requirement intersects with the lower bound function for values of α ranging from 0 to some value less than 1. This means that the bank's advanced measurement approaches (AMA) might be considered reasonable by the regulator. In Case 2, the bank's proposed reserve is lower than the lower bound function for all values of α, giving rise to regulatory concern. In Case 3, the AMA lies entirely above the lower bound, thus generating greater confidence than in Case 1. In all cases, the Standardized Approach would require too much capital reserve

4. Justification of our approach

Our first objective is to find reasonable estimators for the mean and standard deviation of operational risk-related loss for bank b. If we had these, we could use an 'appropriate' standard probability distribution to find the 99.9-th percentile and hence the capital reserve.

The reality is that any individual bank will likely possess only a handful of years of annual loss data and perhaps not even that for all 8×7 categories under Basel II. Such data would be inadequate to estimate means and standard deviations of annual losses with any degree of confidence, leave alone finding the best-fitting standard probability distribution. Therefore, the regulator will have to rely on the pooled operational risk losses reported by many different banks to understand or cross-check the AMA and capital reserve proposed by any bank.

However, banks in the pool vary considerably so the regulator must scale all losses by size of the reporting bank (or that of its pertinent line of business) to get useful statistics. As such, we suggest that the regulator considers all reported losses in the pool as percentages of annual gross income by line of business as well as for the respective bank as a whole. Our underlying assumption is therefore that, other things being equal between two banks, the operational risk losses (actually the parameters of the underlying distribution such as mean and standard deviation) are proportional to the annual gross income for the line of business and for the bank as a whole.

In other words, all the banks' annual operational risk losses – those in the past as well as those in the future – are from the same probability distribution when the losses are percentages of the annual gross income. Therefore, the regulator can average the scaled losses in the pooled losses to estimate the pooled mean loss (μ_p) for a particular line of business or for a bank as a whole expressed as a percentage of annual gross income. Using the Basel Committee's eight lines of business, there would be nine such parameter estimates, eight for the lines of business and one for a bank overall. (Recall that we do not use subscripts or other marks to distinguish between estimates for different lines of business or to distinguish estimates from the unknown parameter values.)

A bank proposing a capital reserve can potentially have a different probability distribution of operational risk losses and therefore a different mean and standard deviation of annual losses than that of the pool. This is because it may have better (or worse) controls on its internal processes than the 'typical' bank in the pool. The regulator can give the benefit of doubt, allowing the bank to report its mean loss for bank b (μ_b) as a percentage of annual gross income from whatever history and/or subjective estimates made by the banks' managers. (Note that the bank's loss models and proposed capital requirement is likely to be based on a combination of its loss database, if any, and the subjective estimates of its managers, but our article is not about the different approaches a bank may take internally.)

On the contrary, the regulator needs to compare the effect of accepting the bank's self-reported expected losses to that of the 'typical' bank in the pool. A simple way for the regulator to do so is to give different relative weights to the two means in producing a combined estimate. The weight can range from 1 down to 0 depending on the level of confidence one has in bank b's own model or in the pooled loss experience.

The regulator can combine the two means as

$$\mu_\alpha = \alpha\mu_b + (1-\alpha)\mu_p \tag{6.3}$$

or equivalently as

$$\mu_\alpha = \alpha(\mu_b - \mu_p) + \mu_p \tag{6.4}$$

The latter form shows α to be the weight given to the difference between the expected loss between bank b and that of the pooled result from all other banks. The regulator does not look for any specific value of α – the motivation is only to determine whether there is any range of α for which the bank's proposed capital requirement exceeds the lower bound our approach provides.

The computation for the standard deviation of loss is slightly different. The very limited dataset for bank b makes it very unreliable to calculate σ_b directly, and subjective estimates may be difficult. *We assume that the coefficient of variation (σ_p/μ_p) calculated across all reporting banks applies to all banks, including the bank that is proposing its capital reserve based on its own AMA.* Multiplied by μ_α, we get an estimate for σ_α, the standard deviation of loss per unit of revenue for bank b.

Therefore, the expected annual losses as percentage of annual gross income for the bank ranges from μ_b to μ_p, and correspondingly, the standard deviation of annual losses ranges from $\sigma_p(\mu_b/\mu_p)$ to σ_p as α goes from 1 down to 0. If we knew the probability distribution, we could determine the 99.9-th percentile using the range of means and standard deviations.

However, we do not know the probability distribution, but we can use standard probability distributions that provide lower values of the 99.9-th percentile and therefore give us a lower bound. We can use different distributions – these will impact how high the lower bound will be and therefore how good (or possibly oversensitive) a bound will be to indicate potential problems relating to a bank underestimating its capital reserve because of its AMA. In the next section, we discuss why we used the lognormal distribution to emphasize that others may similarly find distributions that are even better suited.

5.   Justification for a lower bound using the lognormal distribution

We used simulation parameters based on data from actual banks to see how the 99.9-th percentile of annual losses compares with that of some standard probability distributions. In our simulation, the number of individual losses in any of the 8×7 categories is Poisson-distributed and the loss for each event comes from a lognormal distribution. We assume that all categories are independent with regard to the incidence and size of operational risk losses. The parameters we used for the distributions are based on the summary of operational losses of 89 banks (Basel Committee, 2003).

We assumed the distributions of individual losses per event are all lognormal, with means estimated from average loss per event (mean = total loss in cell/total events in cell). We also computed the standard deviation simply by assuming that a coefficient of variation of 0.5 applied to all means. We made this assumption for our test purposes only because Basel Committee (2003) does not provide raw data that we could use to compute standard deviations. We thus obtain individual loss distributions for each cell. (See Figure 6.2 – the y-axis is different for each cell but the x-axis ranges from 10 000 euros to 10 000 000 euros, truncating distributions from both sides.)

With regard to frequency of losses, we assume the frequency of events follows a Poisson distribution with means based on average number of loss events taken from Basel Committee (2003) loss summary (mean = total loss events in cell/89). (We only need one parameter, the mean, for determining a Poisson distribution.) Thus, we obtain the Poisson distribution of events in a year for each cell. (See Figure 6.3 – the y-axis is different for each cell but the x-axis ranges from 0 loss events to 50 loss events.)

To generate simulated raw data, for each of the 8×7 categories, we simulated the total annual loss by generating a Poisson random variable, N, for the number of loss events. We then summed up N independent and identically distributed lognormal random variables to model the annual loss. We repeated this to produce 10 000-simulated observations for each cell, representing 10 000 years of data to get the total annual loss distribution per cell.

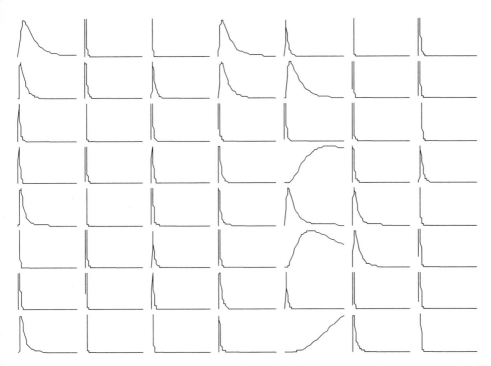

Figure 6.2 Lognormal probability distributions for individual losses given loss event in each cell for eight lines of business and seven categories using parameters estimated from the pooled loss experience of 89 banks summarized in Basel Committee (2003)

We then summed all the losses across the 56 cells for each year, producing 10 000 observations of total annual loss ascribable to operational risk and hence its distribution. (See Figure 6.4 – the x-axis represents losses in euros and the y-axis represents the number of years out of 10 000 where the annual losses are in the range represented by the bar thickness.)

We found the 99.9-th percentile of the simulated annual losses to be 369 million euros (Table 6.2). The average is 78.97 million euros, the standard deviation is 13.386 million, the kurtosis is 12.755 and the skewness is 2.001. We also found that the generated distribution of the total of annual losses across categories has its 99.9-th percentile somewhere between that provided by a lognormal distribution at the lower side and a Pareto distribution on the higher side when all three distributions have the same mean and standard deviation. Therefore, the distribution of simulated losses, formed from a compound Poisson process of lognormal variables, is more skewed and has larger kurtosis than a lognormal distribution for the same mean and standard deviation. Note that this is an empirical observation – we have not derived this result theoretically. Hence, the 99.9-th percentile of the lognormal for the same mean and standard deviation can provide a lower bound on the minimum capital requirement.

We also noted that another standard probability density function, the Pareto distribution, can provide an upper bound function because such a distribution with the same mean and standard deviation has a higher 99.9-th percentile than our simulated loss data. We could use both bounds – a lower such as the lognormal and an upper bound using, for

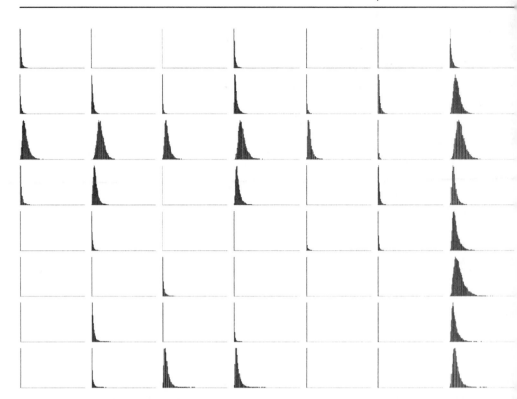

Figure 6.3 Poisson probability distributions for number of loss events in each cell for eight lines of business and seven categories using parameters estimated from the summary of losses provided by the Basel Committee (2003)

instance, the Pareto distribution, but we already have 'upper bounds' from the so-called Basic Indicator and the Standardized Approaches.

We also observed two more things that could form the basis of future work. The first observation was that the 99.9-th percentile for the total losses was almost exactly equal to the square root of the sum of the squares of the losses in each category. We could tie this observation to the fact that the variance of the operational losses in total is equal to the sum across all the 8×7 categories of the variances of the operational losses across category because of the assumption of independence of losses across categories. If we can show this formally, the implication is that each category contributes to the square of the capital reserve requirement with its own capital reserve requirement squared.

Another observation was that taking the log of the simulated annual losses results in a distribution that is still quite skewed but looks more like a lognormal distribution. We took the logarithm of these losses more than once and found that the distribution becomes less skewed each time but remained skewed. This suggests to us that we could use log-lognormal, log-log-lognormal and even log-log-log-lognormal distributions instead of the lognormal distribution to provide a lower bound but this requires further work.

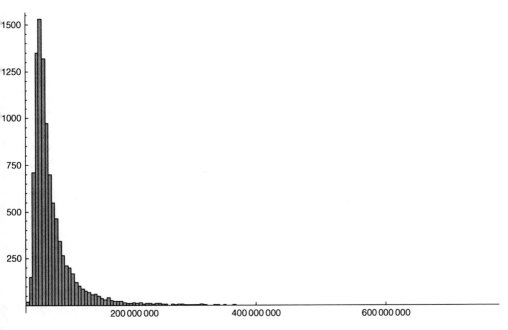

Figure 6.4 The histogram representing the total of annual losses across all the lines of business, the simulation representing annual losses for 10 000 years. Parameters for the simulation were extracted from the pooled losses of 89 banks summarized by the Basel Committee (2003)

Table 6.2 Statistics from simulation results (in millions of euros) for a 'typical' bank along with statistics from lognormal and Pareto distributions that have the same mean and standard deviation as the simulated annual losses. The parameters for the simulation are extracted from the summary of losses (Basel Committee, 2003)

	Lognormal distribution (Lower bound)	Simulated experiment	Pareto distribution (Upper bound)
99.9-th percentile	283.21 m	368.94 m	1315.44 m
Mean	85.91 m		
Standard deviation	37.1 m		
Skewness	1.377	4.263	–
Kurtosis	6.55	37.49	–

6. Conclusion

The Basel II regulation has introduced a broad framework to force banks to address the issue of operational risk measurement but been left sufficiently general to allow different approaches to flourish and thus the field to advance. However, the regulator, and indeed banks themselves, needs a transparent process for assessing whether methodologies employed by individual banks are reasonable. This paper has provided a simple procedure to devise and adopt lower bound functions for the minimum capital requirement to cross-check a particular bank's models. The particular lower bound function we used was based on the lognormal distribution, but other distributions (e.g., log-lognormal) can likely provide more useful bounds.

To the extent banks can buy or otherwise access pooled data of other banks' losses, the procedure also allows banks themselves to cross-check the sophisticated models they may develop in their approach to assessing operational risk. They could also use these bounds with even lower percentiles (say, 90-th percentile) and for quarterly rather than annual losses to check whether their controls are working as expected.

Further work is needed on at least three fronts. First, we need raw pooled operational loss data to refine the simulation experiments. Second, we need to develop theoretical results to underpin our claims and a theory to explain why operational losses might take the chosen distributions for a bank. Finally, we need to find better distributions that could serve as lower bounds for the procedure that we have presented.

References

Alexander, C. (2003). *Operational Risk: Regulation, Analysis and Management*. Financial Times Prentice-Hall.

Basel Committee (2003) *The 2002 Loss Data Collection Exercise for Operational Risk: Summary of the Data Collected. Risk Management Group*. Basel Committee on Banking Supervision, Bank for International Settlements.

Cowell, R.G., Verrall, R.J. and Yoon, Y.K. (forthcoming). Modelling operational risk with Bayesian networks. *Journal of Risk and Insurance*.

Cruz, M. (2003). *Modeling, Measuring and Hedging Operational Risk*. Wiley.

Embrechts, P., Kluppelberg, C. and Mikosch, T. (1997). *Modelling Extremal Events for Insurance and Finance*. Springer.

McNeil, A.J., Frey, R. and Embrechts, P. (2005). *Quantitative Risk Management*. Princeton.

7 Of the credibility of mapping and benchmarking credit risk estimates for internal rating systems

*Vichett Oung**

Abstract

This chapter examines the application of the *credibility theory*, originally developed in the field of risk theory and insurance mathematics, to benchmarking of credit ratings from heterogenous and different credit portfolios. This approach offers an interesting theoretical framework in which robust ratings *comparability* is enabled by explicitly correcting risk bucket estimates for the portfolio's structure effects that are deemed significant. Moreover, statistical tests may then be performed to assess the comparability of whole different risk structures reflected in rating systems. Finally, hierarchical credibility models dealing with nested risk classifications and estimations may also be useful to formalize mapping rules in particular when the benchmark is not of the same granularity. Such 'Credible' mapping could then help build more consistent master scales.

1. Introduction

Both researchers and practitioners have recently been giving growing attention to the issue of benchmarking as a complement and, in some cases, an alternative to the formal validation of internal rating systems based on robust statistical tests. To date and in the current state of the art, the latter still appear to be showing limitations with respect to the extent of their reliability [see for example RTF (2005) and Hamerle *et al.* (2003) for some insights on the subject]. As a result, the recourse to benchmarks is often eventually sought after as the ultimate decision rule in the process of validating rating systems. In this case, a benchmark is given a special credibility, and deviations from its value provide a reason to review or accept the internal estimate at a relatively low cost. Hui *et al.* (2005) argue quite rightly that the benefit of benchmarking lies with its flexibility in the sense that it ultimately gives the experts the scope to select what they consider to be the appropriate benchmark. This will be the case for ratings issued by the so-called ECAI (External Credit Assessment Institutions) eligible for the 'standard approach' under the new Basel II regulation. This credibility is, however, generally implicit and, at best, focused on the observed or assumed quality of the benchmark retained, leaving aside the question of the reliability of the benchmarking process itself, i.e. the consistency of mapping two

* The author is economist at Banque de France, Paris, France
 The views expressed here are his own and do not necessarily represent the views of the Banque de France.

possibly different underlying risk distributions, which, at least for the internal estimates, most likely incorporate *endogenous effects of the portfolio structure*. A prerequisite to robust benchmarking would therefore require some prior inference on the comparability of the internal risk estimate with its benchmark and, ideally, a credibility theory. This chapter examines how insights provided by the *credibility theory* that originated in the field of risk theory and insurance mathematics may pave the way for an application to benchmarking credit risks.

2. Why does the portfolio's structure matter?

2.1. Statistical issues linked with the portfolio risk heterogeneity

The most straightforward concern is with statistical issues. Credit risk is one of the areas in economic and financial empirical research where the paucity of data is most problematic. Default events, relative to the total number of credit loans, remain rare. At a bank portfolio and sub-segment level, it is very unlikely that the biggest corporate risk class accounts for more than a few tens of thousands names. Even in the retail market where portfolios can be expected to be much larger, possibly in the millions, the calibration of high-level confidence default statistics would still require the availability of lengthy series and extensive data. Hence, it appears rather intuitive that the portfolio's structure puts constraints on the availability of data and the accuracy of risk estimates. In this respect, Basel II requires that banks should improve the documentation of default events and the accumulation of statistical data in the future. However, besides data limitation, portfolios' structure also matters for the statistical accuracy of internal rating estimates. Those risks evolve over time, and ratings' *migrations* occur leading to the heterogeneity and instability of data pools underlying the internal rating segmentation. Moreover, rating migration means that following changes in the borrowers' credit quality, internal ratings change or more precisely, borrowers change rating class. This can also be interpreted as shifts in the risk distribution of the bank's portfolio reflecting macroeconomic fluctuations. One could expect that similar macroeconomic changes should be reflected similarly in the risk estimates of two rating systems benchmarked, and in this regard, this is not a concern, unless the perception of these changes by the bank and its rating system are different. Some important insights on this issue are provided by Heiftield, in RTF (2005), who suggests that the dynamics of a bank's rating system lead to very different risk estimates. These are difficult to compare and even back test with regard to the actual defaults observed (e.g. in the case of rating systems that are sensitive to the business cycle). Thus, statistical consistency would require that comparable risk estimates are corrected for the portfolio's heterogeneity.

2.2. Risk transfer mechanisms across the portfolio and double default

Rating migrations also have another economic significance. They mean that risks in general and defaults in particular may be correlated. Standard credit risk models generally capture correlation through the assumed correlation of the latent risk variable (e.g. the distance to default in the case of the Merton or the generalized Vasicek framework) to observable risk factors.

$$Z_{i,t+1} = \omega_i X_{t+1} + \sqrt{1-\omega_i^2} \cdot \varepsilon_{i,t+1}$$

$$Z_{j,t+1} = \omega_j X_{t+1} + \sqrt{1-\omega_j^2} \cdot \varepsilon_{j,t+1}$$

$$\text{cov}\left(Z_{i,t+1}, Z_{j,t+1}\right) = \omega_i \omega_j$$

(7.1)

Equation 7.1 models the potential for all risks in risk bucket i and, more generally, in the portfolio as in Basel II baseline model, to evolve *at the same time*, conditional to the realization of risk Factor X. Default correlation extends, however, to the more general paradigm of the so-called double default correlation. This suggests that besides correlation through common macroeconomic effects, the default (the rating) from one obligor or class of obligors i in the portfolio may consequently trigger the default (the risk deterioration or rating migration) of another obligor or class of obligors j in the same portfolio. One can think more specifically of risk transfer mechanisms as simple as guarantees. As a matter of fact, Basel II does attempt to address the double default correlation that is not explicitly modelled by using a substitution rule according to which the risk of (say) the guarantor is simply substituted onto the risk of the initial obligor. This may potentially underestimate the true risk as the guarantor is usually of a better risk quality than the obligor.

Besides improving modelling, taking into account double default correlation across the portfolio actually gives a better theoretical understanding of portfolio structure effects. Let us now introduce double default correlation between two classes of obligors as a specific risk factor as follows:

$$Z_{i,t+1} = \omega_i X_{t+1} + \sqrt{1-\omega_i^2} \left(\psi_{ij} \theta_{i,t+1} + \sqrt{1-\psi_{ij}^2} \cdot \xi_{i,t+1} \right)$$

$$Z_{j,t+1} = \omega_j X_{t+1} + \sqrt{1-\omega_j^2} \left(\psi_{ij} \theta_{i,t+1} + \sqrt{1-\psi_{ij}^2} \cdot \xi_{j,t+1} \right)$$

$$\text{cov}\left(Z_{i,t+1}, Z_{j,t+1}\right) = \omega_i \omega_j + \psi_{ij}^2 \sqrt{\left(1-\omega_i^2\right)\left(1-\omega_j^2\right)}$$

(7.2)

Equation 7.2 formulates that, conditional to the overall risk factor X, the double correlation between two classes of obligors i and j may occur at any time following the realization of a specific risk factor $\theta_{i,t}$ common to both. This can be generalized to all obligors within the portfolio. Let Θ be the random risk variable taking values $\Theta = \theta_i$, with probability U for all classes of obligors i across the portfolio, Equation 7.2 can be written as

$$Z_{i,t+1} = \omega_i X_{t+1} + \sqrt{1-\omega_i^2} \cdot \left(\psi_{ij} \Theta_{t+1} + \sqrt{1-\psi_{ij}^2} \cdot \xi_{i,t+1} \right),$$

(7.3)

where Θ represents the risk of double default correlation stemming from the portfolio's structure. Modelling double default correlation still remains a difficult theoretical and empirical issue, which is beyond the scope and objective of this chapter. Nevertheless, this analysis suggests that because of such risk transfer mechanisms, the structure of the portfolio implicitly conveys information about potential risk correlation that could impact internal risk estimates at the bucket level. Thus, if two rating systems do not reflect the same portfolio structures, nor the same risk transfer mechanisms, then benchmarking could be spurious. This raises the question of a credibility model for benchmarking. The issue of credibility was extensively studied by researchers on non-life insurance risks. Eventually, a credibility theory was proposed to take into account the potential impact of the portfolio's structure on the sub-class risk estimates.

3. Credible credit ratings and credible credit risk estimates

3.1. Credibility theory and credible ratings

Credibility theory is actually a very old branch of risk theory and non-life insurance mathematics. It was first formulated in the 1920s by American actuaries to correct potential portfolio structure effects on the construction and estimation for risk classes from heterogeneous sub-portfolios. It gained importance and credit with the formalization of its theoretical foundations by Bühlmann (1967) and Bühlmann and Straub (1970) who proposed the standard models of modern credibility theory. The generalization of the approach was intensively reported in the actuarial literature over the ensuing decades, some key references being Hachemeister (1975), Jewell (1975), Sundt (1980) and Norberg (1980) to name a few. The most recent contributions seek to extend the theory to dynamic risk generating processes using Kalman filtering. The Appendix provides some further details on credibility theory of which an overview is introduced for discussion below.

Consider a portfolio of $k = 1, \ldots, N$ risk classes with risk random variable of interest (e.g. number of claims or default events) $Z_{k,i}$ observed for period $i = 1, \ldots, t$ and assumed to be independent. The risk behaviour of each risk class k is assumed to be described by an unknown and unobservable parameter Θ_k. Let Θ be the random variable taking values $\theta = \Theta_1, \Theta_2, \ldots, \Theta_k$ with probability distribution U. Interestingly, actuaries and risk mathematicians name this distribution the portfolio's *structure function* and as a matter of fact U characterizes the heterogeneity of the different risk segment risk estimates in the overall portfolio.

$$\mu(\theta) = E(Z_k/\Theta = \theta)$$

$$m = \int \mu(\theta) dU(\theta) = E(\mu(\Theta))$$

$$a = \int (\mu(\theta) - m)^2 dU(\theta) = V(\mu(\Theta))$$

$$s^2 = E(\sigma^2(\Theta_k)) = \int \sigma^2(\theta) dU(\theta) \quad \text{with} \quad \sigma^2(\Theta) = V(Z_k/\Theta)$$

Provided that all random variables are square integrables in the Hilbert space L^2 generated by $(Z_{k,i})$ and under the above notations, then it can be demonstrated that the best approximation of $\mu(\Theta_k)$, in the sense of the least squares, is defined by its projection on the sub-space A_n of all *affine* combinations of the observations $(1, Z_{k,i})_{1 \leq i \leq t}$. An alternative is a projection on *linear* combinations of observations $(Z_{k,i})_{1\leq i\leq t}$. In the *affine* case the credibility estimator is said to be *non-homogenous* and conversely *homogenous* in the *linear* case (see Appendix). We focus here on the more general case corresponding to the *non-homogenous* estimator. This estimator is given by Bühlmann's famed credibility theorem

$$Cr\left(\mu\left(\theta_k\right)\right)=(1-b_{kt})\,m+b_{kt}\overline{Z}_{k\bullet}.$$

$$\overline{Z}_{k\bullet}=\frac{1}{t}\sum_{k=1}^{N}Z_{ki}\quad\text{and}\quad b_{kt}=\frac{at}{at+s^2}=\frac{t}{t+\dfrac{s^2}{a}}.\tag{7.4}$$

A demonstration of Equation 7.4 is given in Appendix. In its most intuitive and simple form, credibility theorem simply formulates that the optimal estimator of a given risk random variable of interest (say frequencies of claims in the field of insurance risks or frequency of defaults in the field of credit risks) of a sub-portfolio k, conditional to the portfolio's structure, is simply a weighted average between the empirical mean $\overline{Z}_{k\bullet}$ and the estimator obtained on the whole portfolio:

$$Cr\left(E\left(Z_k\right)\right)=(1-b_k)\,E\left(Z\right)+b_k\overline{Z}_k$$

The weights b_k are called *credibility coefficients* and vary between 0 and 1, and $Cr\left(E\left(Z_i\right)\right)$ is also called the *credibility estimator* or *credibility rating*. Thus, in a very simple and intuitive form, this theory suggests that the effects of the portfolio structure should be taken into account in the risk estimate for a risk class k of this portfolio.

3.2. Applying credibility theory to internal risk rating systems

In theory, credibility theory is perfectly applicable to credit risk. In particular, one can note two interesting specific corollary results from the standard credibility theory briefly summarized above:

1. when $a=0$, i.e. $\Theta_1=\Theta_2=\ldots=\Theta_k$ meaning that there is no risk class heterogeneity within the portfolio, then $b_k=0$ and $Cr\left(\mu\left(\Theta_k\right)\right)=m=E\left(\mu\left(\Theta_k\right)\right)$ almost surely;
2. when $s^2>0$ and $a>0$, which is the general case, i.e. there is some risk class heterogeneity within the portfolio. In this case $m=E\left(\mu\left(\Theta_k\right)\right)$ is *probably not* a 'good' estimator of the risk bucket; at the limit $a\rightarrow+\infty$ and $b_k=1$ thus leading to the intuitive conclusion that risk estimates at the portfolio level should almost surely be preferred to the risk bucket level which can be deemed as too unreliable.

These two corollary results may potentially have strong implications for the validation of banks' internal rating systems in the field of credit risks. In particular, corollary 2 actually suggests that in the general case the *structure* of the risk buckets of an internal rating system is likely to reflect the risk heterogeneity of credit portfolios for the reasons mentioned before. A great deal of attention should then be paid to estimating the class' risk estimate. This issue has so far been overlooked by researchers, practitioners and supervisors, whose attention has mainly been devoted to setting a minimum target for the granularity of risk buckets within an internal rating system to maximize its discriminatory power. Taking a different route Hamerle *et al.* (2003) came to a similar conclusion when examining the portfolio's structure effects on the discriminatory power of a rating system using accuracy ratio (AR) type measures. They show in particular that a portfolio's structure effects can lead to *spurious* performance measures. Thus, the accuracy of internal risk rating estimates and performance regardless of the portfolio's structure could be misleading. The following example helps illustrate how credibility theory may be applied to offer a more robust answer to this important issue in the validation of internal rating systems.

Let us now introduce the portfolio structure in the dynamics of the rating system. We consider the default process as the combination of two components:

1. a Poisson distribution that models the number of defaults n for a class of obligors calibrated to a given default level k, conditional on the realization of the parameter θ

$$p_\theta (n = k) = p (n = k \,|\,\theta) = \left(\frac{\theta^k}{k!}\right) \cdot \exp(-\theta) \tag{7.5}$$

with $E(n) = V(n) = \theta$.

2. a Gamma distribution with non-negative scale and shape parameters α, β and density on $[0, +\infty]$, which models the distribution of risks θ within the portfolio. The probability density function is

$$f(\theta) = \frac{\beta^\alpha}{\Gamma(\alpha)} \cdot \theta^{\alpha-1} \cdot \exp(-\beta \cdot \theta) \tag{7.6}$$

with $E(\theta) = \alpha/\beta$ and $V(\theta) = \alpha/\beta^2$.

From Equations 7.5 and 7.6, the number of defaults for any (class of) obligor randomly selected from a portfolio whose structure is characterized by a Gamma distribution with parameters (α, β) has therefore a Gamma–Poisson mixed distribution defined by the probability density function:

$$P(n = k) = \int_0^\infty p(n = k \,|\,\theta) \cdot f(\theta)\, d\theta. \tag{7.7}$$

The calculation in the Appendix shows that Equation 7.7 actually reflects the form of a negative binomial distribution $BN(r, p)$ with parameters $[r = \alpha, p = \beta/(\beta+1)]$ and

$$P(n = k) = \binom{n+r-1}{n} p^r (1-p)^n \tag{7.8}$$

with

$$E(n) = \frac{r(1-p)}{p} = \frac{\alpha}{\beta} \tag{7.9}$$

and

$$V(n) = \frac{k(1-p)}{p^2} = \frac{\alpha(\beta+1)}{\beta^2} \tag{7.10}$$

From the statistical point of view, one can note from Equation 7.10 that the portfolio structure has indeed an *amplifying effect* on the variance of the default process consistent with the economic intuition that double default correlation should bring more uncertainty to risk estimates.

Moreover, the Bühlmann credibility factor within the Gamma–Poisson model has a simple reduced form (see Appendix for calculation):

$$z_k = \frac{(\alpha + N_k)}{(\beta + t)}, \qquad (7.11)$$

where (α, β) are the parameters from the Gamma distribution and N_k is the total number of defaults for class k observed over the period t.

4. An empirical illustration

Table 7.1 simulates the number of defaults observed on a loans' portfolio segmented into 17 risk buckets the size of which is normalized to 100 contracts each. The events of defaults are assumed to be generated by a Poisson process as described by Equation 7.5 and are observed over a 10-year period for the internal rating system studied. In a naïve approach, benchmarking the risk estimates of this rating system to the reference rating scale given by θ^0 would lead to the conclusion that the two rating systems are equivalent.

Assume now that the portfolio structure of this rating system conveys information about default correlation and can be modelled by a Gamma distribution of observable parameters (α, β). Figure 7.1 presents two different cases. Case 1 $(\alpha, \beta) = (3, 1.5)$ simulates a risky portfolio while case 2 $(\alpha, \beta) = (2, 3)$ represents a safer one with more concentrated risks.

Table 7.2 presents the Bühlmann credibility risk estimates given by Equation 7.11 for both case 1 and 2 and compares them to the benchmark rating scale. The results are strikingly different from the naïve approach that concluded in the equivalence of both rating systems. In case 1, the credible estimates suggest that given the portfolio structure and empirical experience, defaults may be already underestimated in low-risk classes,

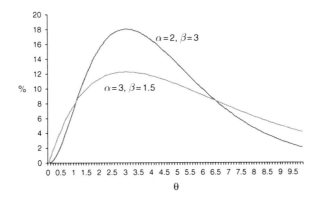

Figure 7.1 Portfolio structure parameters

Table 7.1 Simulated defaults observed

								Risk bucket							
θ^0	0.10%	0.25%	0.50%	0.75%	1.00%	1.50%	2.20%	3.50%	5.00%	7.00%	10.00%	12.00%	15.00%	18.00%	23.00%
Year	1	2	3	4	5	6	7	8	9	10	11	12	13	14	15
1	0	1	0	0	0	1	2	0	8	2	12	15	21	15	30
2	0	0	0	0	2	1	3	2	5	8	10	12	18	19	27
3	0	0	0	0	0	0	3	6	10	9	9	10	12	17	28
4	0	1	0	0	1	0	1	2	4	11	10	13	11	22	18
5	0	1	1	0	0	1	1	4	8	2	14	16	17	15	22
6	0	0	0	0	0	1	4	1	6	5	4	16	13	17	27
7	0	0	2	0	2	2	3	1	9	7	9	15	8	14	24
8	0	1	2	1	0	2	1	4	6	8	8	11	13	11	19
9	1	0	1	3	1	1	4	6	9	8	8	14	13	15	21
10	0	0	0	0	1	0	3	4	6	5	7	11	7	13	33
m	0.1	0.4	0.6	0.4	0.7	0.9	2.5	3	7.1	6.5	9.1	13.3	13.3	15.8	24.9
N	1	4	6	4	7	9	25	30	71	65	91	133	133	158	249

Table 7.2 Benchmarking credible risk estimate

	1	2	3	4	5	6	7	8	9	10	11	12	13	14	15	16	17
Case 1	0.35	0.61	0.78	0.61	0.87	1.04	2.43	2.87	6.43	5.91	8.17	11.83	11.83	14.00	21.91	21.39	25.74
Case 2	0.23	0.46	0.62	0.46	0.69	0.85	2.08	2.46	5.62	5.15	7.15	10.38	10.38	12.31	19.31	18.85	22.69
$\hat{m}$	0.10	0.40	0.60	0.40	0.70	0.90	2.50	3.00	7.10	6.50	9.10	13.30	13.30	15.80	24.90	24.30	29.30
θ^0	0.10	0.25	0.50	0.75	1.00	1.50	2.20	3.50	5.00	7.00	10.00	12.00	15.00	18.00	23.00	26.00	30.00

Table 7.3 Accuracy ratios

	Accuracy ratio
Prior estimates	0.6925
Credible estimates (Case 1)	0.5593
Credible estimates (Case 2)	0.5469

whereas they are conversely overestimated in high-risk classes. This may be due to implicit default correlation and/or wrong calibration. Conversely, portfolio 2 that is concentrated to a far greater extent on good risks gives less credibility to the high-risk estimate.

Overall, the benchmarking of credible risk estimates suggests that the rating systems benchmarked are *not* equivalent. It is interesting to note that credibility estimates also provide a very different view of the discriminative power of the rating system studied. For example, one can observe that credibility estimates for risk buckets 2, 3 and 4 suggest that, conditional to the observations and the portfolio structure effects, these risk class are not discriminative enough, which mitigates the initial impression of a very granular rating scale. The same conclusion can be observed for risk buckets 9 and 10, 12 and 13, as well as 15 and 16. This has two implications, firstly in terms of performance measurement with indicators such as ARs [see RTF (2005) for an extensive presentation of performance indicator for rating systems], and secondly in terms of mapping (see Section 5).

As regards performance measurement, Table 7.3 compares the ARs obtained on prior rating estimates and credible rating estimates for both models (see Appendix for calculation). Thus, one can note that while displaying a rather satisfactory AR close to 0.70 with the prior estimates obtained from Table 7.2, AR calculated for credible estimates from models 1 and 2 are much lower and confirm that the two rating systems are not that discriminative.

Overall, the results illustrated by this case study provide some evidence that naïve benchmarking can lead to spurious inferences and conclusions. On the contrary, it seems that credibility theory applied to credit ratings, and therefore the credible ratings thus obtained may provide some instruments for a more robust benchmarking. Moreover, formal statistical tests again derived from insurance practices such as the Pearson test or the log-rank test may then be performed on credible ratings to complete the approach.

4.1. Statistical tests

A first consequence of potential portfolio structure effects is that rating estimates for risk classes within two different portfolios cannot be compared pair wise, as usually performed in the naïve approach but, conversely, should be benchmarked and tested jointly conditional to the portfolio structure. Furthermore, this means that benchmarking two rating systems should focus instead on assessing the consistency of the respective credible risk rating distributions. Further investigation into the interesting insights of Jewell (1974) with respect to the estimation of whole credible distributions is beyond the scope of this chapter. In effect, however, benchmarking would actually require the systematic comparison of default curves conditional to the portfolio structure. More straightforward statistical tests are, however, available to test the adequacy of the conditional default curves:

- The χ^2 Pearson test is the most familiar approach. It tests the null hypothesis of no significant difference between the two risk estimates benchmarked; it can, however, be rapidly constrained by the need for the same granularity both for rating scales' comparison as well as by the degree of freedom ($N-1$ with N as the number of risk classes).
- An interesting alternative approach is the log-rank test. It resembles the Pearson test but has more extensive power and is often used in life insurance practices to verify the consistency of mortality or survival curves to regulatory benchmarks. In life insurance, the test is used in the temporal dimension to compare two curves $S_A(t)$ and $S_B(t)$ generally representing survival rate (or mortality rate) e. The null hypothesis is the adequacy of the compared curve to the benchmark $H_0 : S_A(t) = S_B(t)$, for $t > 0$. The alternative is that there exists at least one period in time when the survival rates differ H_1: there exists t^* such as $S_A(t^*) \neq S_B(t^*)$.

The log-rank statistics is constructed in the following manner:

$$Q = \underbrace{\frac{(D_A - E_A)^2}{E_A}}_{Q_A} + \underbrace{\frac{(D_B - E_B)^2}{E_B}}_{Q_B}$$

where

$D_A = \sum_t D_{tA}$ is the total number over time of deceases observed for population A.
$D_B = \sum_t D_{tB}$ is the total number over time of deceases observed for population B.
$E_A = \sum_t D_{tA}$ is the total number over time of deceases expected for population A.
$E_B = \sum_t D_{tB}$ is the total number over time of deceases expected for population B.

Under the null hypothesis the statistic Q is then asymptotically distributed according to a χ^2 with one degree of freedom.

One can note that there should be no conceptual difficulty in applying the log-rank test cross-section across risk class k of two given rating systems A and B to compare the corresponding default curves $S_A(k)$ and $S_B(k)$. We construct the log-rank test for our example following results displayed in Table 7.2. In this example, both rating systems have been calibrated to have ex ante the same expected default rates. Recall that according to Equation 7.5, $E(n) = V(n) = \theta$. Thus, obtaining the log-rank statistic from Table 7.2 is straightforward. The value of the log-rank Q statistic for benchmarking both rating systems is 15.91, well above the critical value of 3.84 at the 5% confidence level, thus rejecting the null hypothesis of equivalence of the two rating systems benchmarked (Table 7.4).

Table 7.4 Log-rank test

	1	2	3	4	5	6	7	8	9	10	11	12	13	14	15	16	17	Total
Case	10.35	0.61	0.78	0.61	0.87	1.04	2.43	2.87	6.43	5.91	8.17	11.83	11.83	14.00	21.91	21.39	25.74	
Case	20.23	0.46	0.62	0.46	0.69	0.85	2.08	2.46	5.62	5.15	7.15	10.38	10.38	12.31	19.31	18.85	22.69	
$\tilde{m}$	0.10	0.40	0.60	0.40	0.70	0.90	2.50	3.00	7.10	6.50	9.10	13.30	13.30	15.80	24.90	24.30	29.30	
θ^0	0.10	0.25	0.50	0.75	1.00	1.50	2.20	3.50	5.00	7.00	10.00	12.00	15.00	18.00	23.00	26.00	30.00	
Q_A	0.63	0.52	0.16	0.03	0.02	0.14	0.02	0.11	0.41	0.17	0.33	0.00	0.67	0.89	0.05	0.82	0.60	5.57
Q_B	0.17	0.18	0.03	0.11	0.10	0.28	0.01	0.31	0.08	0.49	0.81	0.22	1.42	1.80	0.59	1.97	1.78	10.34

5. Credible mapping

5.1. The mapping process

We define mapping as the process that normalizes the granularity of two rating scales with different numbers of buckets. Mapping is thus an important preliminary step:

- to allow the benchmarking of the rating system studied to a reference external rating scale (S&P's, Moody's, etc.) with a different granularity and
- also to enable the prior aggregation of sub-ratings systems on a master scale, which is then used for benchmarking to external references.

The latter specific issue is particularly important for internal rating systems implemented by banks and the credit portfolios of which are usually segmented according to refined risk and economic characteristics reflecting business practices tailored to address specific and very diverse market segments. As noted in RTF (2005), internal rating systems are usually designed in a 'bottom-up' approach, better reflecting operational reality, but rating grades are often ultimately *mapped* onto a *master scale* that is then calibrated or compared to an external benchmark.

Notwithstanding the effects of this segmentation on the accuracy of risk bucket estimates previously mentioned, internal rating systems are ultimately facing an aggregation consistency issue. Thus, as for credible ratings discussed above, credible benchmarking would also require credible mapping. In this regard, generalizations and extensions of the standard credibility theory, such as those initially proposed by Jewell (1975) for segmented insurance portfolios, may actually also provide some theoretical background to credible mapping in the case of credit ratings (Figure 7.2).

The insights again issue from the literature on credibility for insurance risks. We introduce here the basic principles of hierarchical credibility as first introduced by Jewell (1975). More recent developments can be found in particular in Dannenburg *et al.* (1996) and Ohlsson (2006), who proposes an interesting multiplicative hierarchical credibility model.

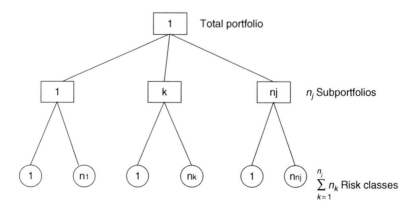

Figure 7.2 Jewell's (1975) hierarchical credibility model

Consider a portfolio segmented in n sub-portfolios that may be risk classes. Each bucket can itself be segmented and so on so that the overall portfolio is hierarchically segmented in j levels of aggregation. One can think, for example, of a bank portfolio segmented by country sub-portfolios that are in turn segmented by market (retail, corporate, SME, etc.) and again by sector or products (industry, services, mortgage, etc.). Ultimately, each jth level sub-portfolio is segmented in n_{nj} risk buckets.

Following Dannenburg et al. (1996), we use the triplet (i, j, k) to characterize any hierarchical structure of a portfolio of contracts, in this case loans, and more precisely risk bucket k of sub-portfolio j at the ith level of sub-aggregation.

The observations are realizations of random variables $(Z_{i,j,k})$ that are supposed to be square integrables. The principles of credibility modelling presented above are applied to the hierarchical structure: at each given level i each aggregate sub-portfolio or new cluster j is assumed to be characterized by a latent risk parameter θ_j, which is not observable and also assumed to be a random variable. In addition, each risk class level k of sub-portfolio j is also assumed to be characterized by an unobservable risk parameter θ_{jk}. One then obtains the following specifications:

- at the overall portfolio level: $E(Z_{i,j,k}) = m$;
- at the jth sub-portfolio in level i: $E(Z_{i,j,k}|\theta_j) = \mu_0(\theta_j)$ and
- at the risk class k of the jth aggregate in level i: $E(Z_{i,j,k}|\theta_j, \theta_{jk}) = \mu(\theta_j, \theta_{jk})$.

Furthermore, let

$$b = \text{var}(\mu_0(\theta_j)),$$
$$a = E(\text{var}(\mu(\theta_j, \theta_{jk})|\theta_j)) \text{ and}$$
$$s^2 = E(\sigma^2(\mu(\theta_j, \theta_{jk}))).$$

It can then be shown that credibility estimates obtained for any aggregate class j are given by

$$Cr(E(Z_j)) = (1 - q_j)m + q_j \overline{Z_{j\bullet\bullet}}^{zw}, \tag{7.12}$$

where

$$q_j = \frac{z_{j\bullet}}{z_{j\bullet} + \dfrac{a}{b}} \tag{7.13}$$

is the credibility factor for aggregate class j;

$$z_{j\bullet} = \sum_k z_{jk} \tag{7.14}$$

is the aggregate credibility estimate for all classes of aggregate class j with

$$z_{jk} = \frac{w_{jk\bullet}}{w_{jk\bullet} + \dfrac{s^2}{a}} \tag{7.15}$$

as credibility factor for sub-class k; $w_{jk\bullet} = \sum_i w_{i,j,k}$ is the total aggregate number of observations at level i for aggregate class j and sub-class k.

The credibility factor for sub-class k here differs slightly from the standard credibility factor in Equation 7.4 by the introduction of 'weights', reflected by the relative size of sub-segments prior to aggregation.

$$\overline{Z_{jk\bullet}}^{w} = \frac{\sum_i w_{ijk} Z_{i,j,k}}{\sum_i w_{ijk}} \tag{7.16}$$

if the average defaults for sub-portfolio j and class k weighted by total aggregated number of exposures at level i.

$$\overline{Z_{j\bullet\bullet}}^{zw} = \frac{\sum_k z_{jk} \overline{Z_{jk\bullet}}^{w}}{\sum_k z_{jk}} \tag{7.17}$$

is the average *credible* risk estimate for aggregate class j weighted by credibility factors of sub-classes k.

5.2. An empirical illustration

Assume that the following rating scale with 10 grades is a reference scale (Table 7.5).

One may wish to map the initial 17 grades onto an equivalent 10 grades master scale, and assume that all prerequisites in terms of mapping consistency are fulfilled (same type of portfolio). Often a rather simple approach is used, by which a 'naïve mapping' is directly performed by clustering risk buckets together such as in Table 7.6. In this case, one would assume that, based on prior risk estimates, bucket 1 of the reference rating scale, which displays a default rate of 0.25%, would map to buckets 1 and 2 with respective prior estimates of 0.1% and 0.25%.

However, a 'credible' mapping based on credible estimates would suggest a different approach. Firstly, the previous results obtained from risk class credibility estimates for

Table 7.5 Reference benchmark scale

Bucket	1	2	3	4	5	6	7	8	9	10
% Defaults	0.25	0.75	1	2	5	7	10	15	25	30

Table 7.6 Bucket mapping

Benchmark	1	2	3	4	5	6	7	8	9	10
Naïve mapping	1 + 2	3 + 4	5 + 6	7	8 + 9	10	11 + 12	13 + 14	15 + 16	17
Credible mapping	1	2 + 3 + 4	5 + 6	7	8	9 + 10 + 11	12 + 13 + 14	15 + 16	17	

model 1 (see Table 7.2) indicated a lack of discriminative power of a number of buckets. A first stage in a credible mapping would then require these buckets to be clustered together as presented in Table 7.6. Secondly, proper credible risk estimates for the mapping performed would need to be calculated. This is undoubtedly more satisfactory than in the naïve approach. To do so, we apply Jewell's two-level credibility model to model 1 previously studied: in our simple case study, all classes have the same weight and the same credibility factor, thus greatly simplifying the calculation that can then be directly obtained from previous results from Table 7.4. As a matter of fact, the calculation reduces to a second-step estimation of standard credibility factor on the resulting 10 buckets mapped rating scale and corresponding portfolio. As before, the specific Gamma–Poisson model underlying model 1 allows the use of Equation 7.11 to directly derive new credibility estimates for the mapped rating scale. Table 7.7a presents the data used.

The resulting mapped rating scale and corresponding credible estimates can now be benchmarked with the reference rating scale. The results presented in Table 7.7b appear interesting and quite satisfactory. Most of the buckets compared have rather similar risk estimates. Some larger discrepancies are observed, however, noticeably for 10. The log-rank test gives nevertheless the value 2.51 below the 3.84 critical value, thus accepting the null hypothesis of equivalence between the two rating scales. Moreover, the credible mapping, i.e. mapping performed on credible estimates, also resulted in an increased discriminative power, from initially 0.5593 (see Table 7.3) to 0.6121.

Table 7.7a Defaults (%) for mapped rating scale

Year	1	2 + 3 + 4	5 + 6	7	8	9 + 10 + 11	12 + 13	14	15 + 16	17
1	0.00	0.33	0.50	2.00	0.00	7.33	18.00	15.00	26.00	24.00
2	0.00	0.00	1.50	3.00	2.00	7.67	15.00	19.00	29.00	41.00
3	0.00	0.00	0.00	3.00	6.00	9.33	11.00	17.00	28.00	27.00
4	0.00	0.33	0.50	1.00	2.00	8.33	12.00	22.00	20.00	26.00
5	0.00	0.67	0.50	1.00	4.00	8.00	16.50	15.00	19.00	29.00
6	0.00	0.00	0.50	4.00	1.00	5.00	14.50	17.00	24.00	28.00
7	0.00	0.67	2.00	3.00	1.00	8.33	11.50	14.00	22.50	35.00
8	0.00	1.33	1.00	1.00	4.00	7.33	12.00	11.00	25.00	26.00
9	1.00	1.33	1.00	4.00	6.00	8.33	13.50	15.00	23.00	28.00
10	0.00	0.00	0.50	3.00	4.00	6.00	9.00	13.00	29.50	29.00
Average	0.10	0.47	0.80	2.50	3.00	7.57	13.30	15.80	24.60	29.30

Table 7.7b Credible estimates for mapped rating scale

Benchmark	1	2	3	4	5	6	7	8	9	10
Risk estimate	0.25	0.75	1.00	2.00	5.00	7.00	10.00	15.00	25.00	30.00
Credible mapping	1	2 + 3 + 4	5 + 6	7	8	9 + 10 + 11	12 + 13 + 14	15 + 16	17	
Credible estimate	0.35	0.67	0.96	2.43	2.87	6.84	11.83	14.00	21.65	25.74

6. Conclusions

Benchmarking will take on a far greater importance for internal credit rating validation processes if it can be formalized into a robust approach. In particular, the benchmarking process needs to be given some formal *credibility*. To this end, this chapter has looked at the *credibility theory* initially developed in the field of risk theory for non-life insurance claims' portfolios. The conclusions reached in this chapter are that in many respects, credibility theory may actually be applicable to credit portfolios. In particular

- Risk estimates for internal rating systems may be affected by the *portfolio structure* that reflects not only statistical calibration problems but also economic phenomenon such as double default correlation. Therefore, benchmarking may become problematic if the ratings compared on the basis of their estimates are biased or include idiosyncratic portfolio structure effects. Credibility theory was developed to address such concerns. What makes credibility theory particularly attractive when applied to benchmarking credit rating and credit risk estimates is that it offers an elegant theoretical framework that enables ratings *comparability* by explicitly correcting risk bucket estimates for the portfolio's structure effects.
- Because of the portfolios' structure effects, benchmarking credible risk estimates actually broadens the issue from benchmarking *ratings* to benchmarking the *rating system's structure*, which ultimately is a more satisfactory and consistent approach. Moreover, statistical tests may then be performed to assess the comparability of different risk structures.
- Developments in credibility theory such as hierarchical credibility models may also be useful to formalize mapping rules in particular for using *master scales*. This is needed when the benchmark is not of the same granularity. 'Credible' mapping could help to build more consistent master scales.

7. Acknowledgements

The author thanks Christian Robert (CREST) and Alain Monfort (CNAM) for inspiring many of the insights developed in this work.

References

Bühlmann, H. (1967) Experience rating and credibility. *Astin Bulletin*, 4, 199–207.
Bühlmann, H. and Straub, E. (1970) Glaudwürdigkeit für Schadensätze. *Bulletin de l'Association Suisse des Actuaires*, 70, 117–47.
Dannenburg, D.R., Kaas R. and Goovaerts, M.J. (1996) *Practical Actuarial Credibility Models*. Amsterdam: IAE.
Hui, C.H., Wong, T.C., Lo, C.F., *et al.* (2005) Benchmarking model of default probabilities of listed companies. *The Journal of Fixed Income*, 3, September, pp. 76–86.
Hachemeister, C.A. (1975) Credibility for regression models with application to trend. In *Credibility, Theory and Application*. Proceedings of the Berkeley Actuarial Research Conference on Credibility. New York: Academic Press, pp. 129–63.

Hamerle, A., Rauhmeier, R. and Rösch, D. (2003) *Uses and Misuses of Measures for Credit Rating Accuracy.* Working Paper, April. Regensburg, Germany: Department of Statistics, Faculty of Business and Economics, University of Regensburg.

Jewell, W.S. (1974) The credible distribution. *Astin Bulletin*, 7, 237–69.

Jewell, W.S. (1975) Regularity conditions for exact credibility. *Astin Bulletin*, 8, 336–41.

Norberg, R. (1980) The credibility approach to experience rating. *Scandinavian Actuarial Journal*, 1, 73–80.

Norberg, R. (1996) Hierarchical credibility: analysis of a random effect linear model with nested classification. *Scandinavian Actuarial Journal*, 1, 204–22.

Research Task Force (RTF) (2005) *Studies on the Validation of Internal Rating Systems*, Basel Committee working paper, 14 May.

Ohlsson, E. (2006) *Credibility Estimators in Multiplicative Models*, Research report 3. Mathematical Statistics, Stockholm University.

Sundt, B. (1980) A multi-level hierarchical credibility regression model. *Scandinavian Actuarial Journal*, 1, 25–32.

Appendix

1. Further elements of modern credibility theory

Although credibility concepts originated in the early 1900s, the modern credibility theory was formally established by Bühlmann (1967), who gave it robust mathematical foundations. Recall the fundamental relation of credibility theory given by Equation 7.4

$$Cr\left(\mu\left(\theta_k\right)\right) = \left(1 - b_{kt}\right)m + b_{kt}\overline{Z}_{k\bullet}.$$

The initial idea of credibility theory was to project $\mu\left(\theta\right)$ on sub-spaces generated by $(1, Z_{k,i})$. The resulting non-homogenous credibility estimator is perfectly sound on the theoretical level but can be difficult to use at a practical level, as the required conditional expectation $\mu\left(\theta\right) = E\left(Z_k / \Theta = \theta\right)$ is not always known ex ante. In particular, it necessitates prior assumptions on the form of the *portfolio's structure* function U, which characterizes the probability distribution of risk parameter θ across the portfolio. Under some specific assumptions, however, the problem can be alleviated as demonstrated by Jewell (1975) and an exact solution obtained.

Alternatively, if one does not wish to make such assumptions, an empirical credibility estimator can still be obtained by restricting the problem to a projection on $(Z_{k,i})$, sub-space of *linear combinations* of the observations. In theory, the resulting projection, which is a random variable, does not have same mean as $\mu\left(\theta\right)$ and thus does not comply with the required property. To alleviate this, a best estimate can be obtained by imposing a constraint: the best estimate of the credibility factor, said to be homogenous, is then obtained by the combination of linear combinations of observations that satisfies at the same time the constraint $E\left(Cr^h\left(\mu\left(\theta_k\right)\right)\right) = E\left(\mu\left(\theta_k\right)\right)$. The estimation error is, however, larger.

2. Proof of the credibility fundamental relation

We recall here the initial demonstration carried by Bühlmann (1967) deriving the credibility relation in the *non-homogenous* case. Alternative proofs and generalizations to the hierarchical credibility framework can be found in Jewell (1975), Sundt (1980) and Norberg (1996), among others.

Consider a portfolio of $k = 1, \ldots, n$ risk classes with risk random variable of interest (e.g. number of claims or default events) $Z_{k,i}$ observed for period $i = 1, \ldots, t$ and assumed to be independent. The risk behaviour of each risk class k is assumed to be described by an unknown and unobservable parameter Θ_k. Let Θ be the random variable taking values $\theta = \Theta_1, \Theta_2, \ldots, \Theta_k$ with probability distribution U.

Notations

Let

$\mu(\theta) = E(Z_k/\Theta_k = \theta)$, the expected risk for class k conditional to risk driver θ.

$m = \int \mu(\theta) \, dU(\theta) = E(\mu(\Theta))$, the expected risk over the whole portfolio.

$a = \int (\mu(\theta) - m)^2 \, dU(\theta) = V(\mu(\Theta))$, the variance of risks across the whole portfolio.

$s^2 = E(\sigma^2(\Theta_k)) = \int \sigma^2(\theta) \, dU(\theta)$ with $\sigma^2(\Theta) = V(Z_k/\Theta_k)$, the variance over time of empirical risk estimate Z_k for class k, conditional to risk driver θ_k.

Assumptions

1. $\forall k = 1, \ldots, n; \forall j \geq 1$, the random variables Z_{kj} are square integrables with $E(Z_{kj}|\Theta_k) = \mu(\Theta_k)$ independent from j and $\mathrm{cov}(Z_{kj}, Z_{ki}|\Theta_k) = \sigma^2(\Theta_k)\delta_{ij}$ with $\delta ij = \begin{cases} 1 \text{ if } i = j \\ 0 \text{ if } i \neq j \end{cases}$.

2. $\forall k = 1, \ldots, n$ and $p = 1, \ldots, n$ with $p \neq n$, the random vectors $(\Theta_k, Z_{ki} : i \geq 1)$ and $(\Theta_p, Z_{pj} : j \geq 1)$ are independent in probability, and the random variables $\Theta_1, \Theta_2, \ldots, \Theta_k$ are independent and have the same probability law. Assumption 2 formulates that, conditional to the risk driver of a given class, each (loan) contract is independent.

Theorem

Under assumptions 1 and 2, the best linear (non-homogenous) (Bühlmann, 1967) estimator of the expected risk for class k conditional to risk driver θ is

$$Cr(\mu(\Theta_k)) = (1 - b_{kt})m + b_{kt}\overline{Z}_{k\bullet},$$

where

$$\overline{Z}_{k\bullet} = \frac{1}{t}\sum_{k=1}^{N} Z_{ki} \text{ mean of empirical estimates for class k over time}$$

and $\quad b_{kt} = \dfrac{at}{at + s^2} = \dfrac{t}{t + \frac{s^2}{a}}.$

Proof   One seeks the best linear approximation of $\mu(\Theta_k)$, the risk of class k with risk driver Θ_k, by a random variable $W = Cr(\mu(\Theta_k))$ that is assumed to be an affine combination of all the observations Z_{ij} of the whole portfolio.

$$W = c_0 + \sum_{i=1}^{n}\sum_{j=1}^{t} c_{ij} Z_{ij}$$

W is therefore the orthogonal projection of $\mu(\Theta_k)$ on the sub-space spanned by the random variables $(1, Z_{ij})$. The vector $(\mu(\Theta_k) - W)$ is then orthogonal to this sub-space and the following $nt + 1$ relations are satisfied (scalar product):

$$E((\mu(\Theta) - W) \cdot 1) = 0$$

$$E((\mu(\Theta) - W) \cdot Z_{pl}) = 0 \quad \text{for} \quad p = 1, \ldots, n \quad \text{and} \quad l = 1, \ldots, t$$

From the first relation, one gets $m = c_0 + \sum\limits_{i=1}^{n}\sum\limits_{j=1}^{t} c_{ij} m$

The other relations are equivalent to $\text{cov}(\mu(\theta_k), Z_{pl}) = \text{cov}(W, Z_{pl})$

• $p = k$ then for each $t = 1, \ldots, t$

$$a = a\sum_{i=1}^{n} c_{ki} + s^2 \sum_{i=1}^{t} c_{ki}\delta_{il}, \quad \text{where} \quad \delta_{il} = \begin{cases} 1 \text{ if } i = 1 \\ 0 \text{ if } i \neq 1 \end{cases}$$

$$a\left(1 - a\sum_{i=1}^{t} c_{ki}\right) = s^2 c_{ki},$$

thus, $c_{k1} = c_{k2} = \ldots = c_{kt} = \dfrac{a}{at + s^2}$

• $p \neq k$
Under assumption 2, one gets

$$0 = \sum_{i=1}^{n}\sum_{j=1}^{t} c_{ij}\,\text{cov}(Z_{ij}, Z_{pl})$$

$$0 = \sum_{j=1}^{t} c_{pj}(a + \delta_{jl}s^2) = \sum_{j=1}^{t} c_{pj}a + c_{pl}s^2, \quad \text{thus} \quad c_{p1} = \ldots = c_{pt} = 0$$

Overall, the solution reduces to

$$W = c_0 + c_1\sum_{j=1}^{t} Z_{kj}, \quad \text{with} \quad c_0 = \left(1 - \frac{at}{at + s^2}\right)m \quad \text{and} \quad c_1 = \frac{a}{at + s^2},$$

which demonstrates the results of Equation 7.4.

3. Mixed Gamma–Poisson distribution and negative binomial

Conditional to θ, the probability to observe k defaults is Poisson and given by

$$p(n = k|\theta) = \theta^k \frac{e^{-\theta}}{k!}$$

However, θ is a random variable with Gamma probability density

$$f(\theta) = \frac{\beta^\alpha}{\Gamma(\alpha)} \cdot \theta^{\alpha-1} \cdot \exp(-\beta \cdot \theta).$$

The probability to observe k defaults according to a Gamma–Poisson process is therefore

$$g(n = k) = \int_0^{+\infty} p(n = k|\theta)f(\theta)\,d\theta$$

$$= \int_0^{+\infty} \theta^k \frac{e^{-\theta}}{k!} \frac{\beta^\alpha}{\Gamma(\alpha)}\theta^{\alpha-1}e^{-\beta\theta}\,d\theta$$

$$= \frac{\beta^\alpha}{k!\Gamma(\alpha)} \int_0^{+\infty} \theta^{k+\alpha-1} e^{\theta(\beta-1)}\,d\theta$$

$$= \frac{\beta^{\alpha}}{k!\Gamma(\alpha)} \Gamma(k+\alpha)(\beta+1)^{-(k+\alpha)}$$

$$= \left\{ \frac{\Gamma(k+\alpha)}{k!\Gamma(\alpha)} \right\} \left\{ \frac{\beta}{\beta+1} \right\}^{\alpha} \left\{ \frac{1}{\beta+1} \right\}^{k}$$

or

$$g(n=k) = \left\{ \frac{(k+\alpha-1)!}{(\alpha-1)!k!} \right\} \left\{ \frac{\beta}{\beta+1} \right\}^{\alpha} \left\{ 1 - \frac{\beta}{\beta+1} \right\}^{k}.$$

Let $p = \frac{\beta}{\beta+1}$, then

$$g(n=k) = \binom{k+\alpha-1}{k} p^{\alpha}(1-p)^{k},$$

which is the form of a negative binomial $\left(\alpha, \frac{\beta}{1+\beta}\right)$ with expectation α/β and variance $\frac{\alpha(\beta+1)}{\beta^2} > \frac{\alpha}{\beta^2}$.

4. Calculation of the Bühlmann credibility estimate under the Gamma–Poisson model

θ follows a Gamma distribution with parameters (α, β).
  Recall from Equation 7.4 that

$$Cr(\mu(\theta_k)) = (1-b_{kt})m + b_{kt}\overline{Z}_{k\bullet}$$

$$\overline{Z}_{k\bullet} = \frac{1}{t}\sum_{k=1}^{N} Z_{ki} \quad \text{and} \quad b_{kt} = \frac{at}{at+s^2} = \frac{t}{t+\frac{s^2}{a}}$$

with

$$a = \int (\mu(\theta) - m)^2 dU(\theta) = V(\mu(\Theta)) = \frac{\alpha}{\beta^2}$$

$$s^2 = E(\sigma^2(\Theta_k)) = \int \sigma^2(\theta)dU(\theta) = \frac{\alpha}{\beta}.$$

N_k defaults over t years. The observed annual default frequency is N_k/t
  The credibility of the t years of observations is

$$b_{kt} = \frac{t}{t+\frac{s^2}{a}} = \frac{t}{t+\frac{\alpha/\beta}{\alpha/\beta^2}} = \frac{t}{t+\beta}$$

The ex ante hypothesis of defaults frequencies at the portfolio level is $m = E(\theta) = \frac{\alpha}{\beta}$

Thus the credible estimate for class k is

$$Cr\left(\mu\left(\theta_k\right)\right) = (1 - b_{kt})\,m + b_{kt}\overline{Z}_{k\bullet} = \left(1 - \frac{t}{t+\beta}\right)\cdot\frac{\alpha}{\beta} + \frac{t}{t+\beta}\cdot\frac{N_k}{t} = \frac{\alpha + N_k}{\beta + t}.$$

5. Calculation of accuracy ratio

The demonstration can be found in Hamerle *et al.* (2003):

$$AR = \frac{1}{1 - \bar{\lambda}}\cdot Gini,$$

$\bar{\lambda}$ is the average default rate, with Gini coefficient defined as

$$Gini = 1 + \frac{1}{N} - \frac{2}{N^2\bar{\lambda}}\left[N\lambda_1 + (N-1)\lambda_2 + \ldots + \lambda_N\right],$$

where N is the total number of buckets and λ_i is the default rate for risk bucket i.

8 Analytic models of the ROC Curve

Applications to credit rating model validation

Stephen Satchell and Wei Xia[†][‡]

Abstract

In this chapter, the authors use the concept of the population receiver operating characteristic (ROC) curve to build analytic models of ROC curves. Information about the population properties can be used to gain greater accuracy of estimation relative to the non-parametric methods currently in vogue. If used properly, this is particularly helpful in some situations where the number of sick loans is rather small, a situation frequently met in practice and in periods of benign macro-economic background.

1. Introduction

Following the publication of the 'International Convergence of Capital Measurement and Capital Standards: A Revised Framework' by the Basel Committee on Banking Supervision (BCBS) (known as Basel II) in June 2004, qualified banks are now allowed to use the internal rating-based (IRB) credit risk modelling approach for risk modelling and economical capital calculation. One of the important components of most IRB risk models is the rating system used for transforming and assigning the probability of default (PD) to each obligor in the credit portfolio, and over the last three decades, banks and public ratings agencies have developed a variety of rating methodologies. Therefore, questions arise as to which of these methodologies deliver acceptable discriminatory power between the defaulting and non-defaulting obligor ex ante and which methodologies would be preferred for different obligor sub-groups. It has become increasingly important for both the regulator and the banks to quantify and judge the quality of rating systems.

This concern is reflected and stressed in the recent BCBS working paper No.14 (2005), which summarizes a number of statistical methodologies for assessing discriminatory power described in the literature, for example, cumulative accuracy profile (CAP), receiver operating characteristic (ROC), Bayesian error rate, conditional information entropy ratio (CIER), Kendall's τ and Somers' D, Brier score, inter alia. Among those methodologies, the most popular ones are CAP and its summary index, the accuracy ratio (AR), as well as ROC and its summary index known as the area under the ROC (AUROC) curve. It is worth noting that, unlike some other measures that do not take sample size into account and are therefore

* Faculty of Economics, University of Cambridge and Trinity College, Cambridge, UK
† Doctoral Student, Birkbeck College, University of London, UK
‡ Visiting Lecturer, University of International Business and Economics, Beijing, China

substantially affected by statistical errors, the CAP and the ROC measures explicitly account for the size of the default sample and, thus, can be used for direct rating model comparison.

A detailed explanation of the CAP is presented in Sobehart *et al.* (2000) and Sobehart and Keenan (2004). ROC has long been used in medicine, psychology and signal detection theory, so there is a large body of literature that analyses the properties of the ROC curve. Bamber (1975) shows that the AUROC is related to the Mann–Whitney statistic and also discusses several different methods for constructing confidence intervals (CIs). An overview of possible applications of the ROC curves is given by Swets (1988). Sobehart and Keenan (2001) introduce the ROC concept to internal rating model validation and focus on the calculation and interpretation of the ROC measure. Engelmann *et al.* (2003) show that AR is a linear transformation of AUROC; their work complements the work of Sobehart and Keenan (2001) with more statistical analysis of the ROC. However, the previous work with which we are familiar has used a finite sample of empirical or simulated data, but no one has analysed the analytic properties of the ROC curve and the ROC measure under parametric assumptions for the distribution of the rating scores.

In this chapter, we further explore the statistical properties of the ROC curve and its summary indices, especially under a number of rating score distribution assumptions. We focus on the analytical properties of the ROC curve alone as the CAP measure is just a linear transformation of the ROC measure.

In Section 2, to keep this chapter self-contained, we briefly introduce the credit rating model validation background and explain the concepts and definitions of ROC and CAP. A general equation for the ROC curve is derived. By assuming the existence of probability density functions (PDFs) of the two variables that construct the ROC curve, an unrestrictive assumption, we show that there is a link between the first derivative of the curve and the likelihood ratio (LR) of the two variables, a result derived by different methods in Bamber (1975).

In Section 3, by further assuming certain statistical distributions for the credit rating scores, we derive analytic solutions for the ROC curve and its summary indices. In particular, when the underlying distributions are both negative exponential distributions, we have a closed-form solution.

In Section 4, we apply the results derived in Section 3 to simulated data. Performance evaluation reports are presented. Section 5 concludes.

We find that estimation results from our analytic approach are as good as and, in many cases, better than the non-parametric AUROC ones. The accuracy of our approach is limited by the continuous rating score assumption and also affected by the accuracy of estimation of the distribution parameters on rating score samples in some cases. However, it offers direct insight into more complex situations and is, we argue, a better tool in credit rating model selection procedure as the analytic solution can be used as an objective function.

2. Theoretical implications and applications

In this section, we first briefly review the credit rating system methodology, in particular the CAP and the ROC measures. The content presented in this part is very similar to Engelmann *et al.* (2003) and BCBS working paper No.14 (2005). Then, we introduce the ordinal dominance graph (ODG), where ROC is a special case of ODG and some interesting theoretical implications of the ROC curve will be given.

2.1. The validation of credit rating system

The statistical analysis of rating models is based on the assumption that for a predefined time horizon, there are two groups of bank obligors: those who will be in default, called defaulters, and those who will not be in default, called non-defaulters. It is not observable in advance whether an obligor will be a defaulter or a non-defaulter in the next time horizon. Banks have loan books or credit portfolios; they have to assess an obligor's future status based on a set of his or her present observable characteristics. Rating systems may be regarded as classification tools to provide signals and indications of the obligor's possible future status. A rating score is returned for each obligor based on a rating model, usually an expert judgment model. The main principal of rating systems is that 'the better a grade, the smaller the proportion of defaulters and the greater the proportion of non-defaulters who are assigned this grade'. Some quantitative examples are the famous Altman's Z score or some scores from a Logit model.

Therefore, the quality of a rating system is determined by its discriminatory power between non-defaulting obligors and defaulters ex ante for a specific time horizon, usually a year. The CAP measure and ROC provide statistical measures to assess the discriminatory power of various rating models based on historical (ex post) data.

2.2. CAP and AR

Consider an arbitrary rating model that produces a rating score, where a high score is usually an indicator of a low default probability. To obtain the CAP curve, all debtors are first ordered by their respective scores, from riskiest to safest, i.e. from the debtor with the lowest score to the debtor with the highest score. For a given fraction x of the total number of debtors, the CAP curve is constructed by calculating the percentage $d(x)$ of the defaulters whose rating scores are equal to, or lower than, the maximum score of fraction x. This is done for x ranging from 0% to 100%. Figure 8.1 illustrates CAP curves.

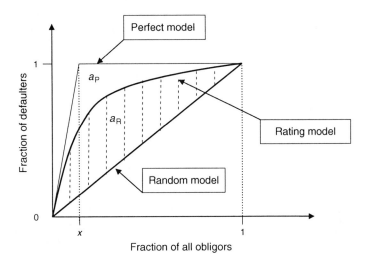

Figure 8.1 Cumulative accuracy profile

A perfect rating model will assign the lowest scores to the defaulters. In this case the CAP increases linearly to one, then remains constant. For a random model withou any discriminative power, the fraction x of all debtors with the lowest rating scores wi contain x percent of all defaulters. The real rating system lies somewhere in between thes two extremes. The quality of a rating system is measured by the AR. It is defined as th ratio of the area a_R between the CAP of the rating model being validated and the CA of the random model and the area a_P between the CAP of the perfect rating model an the CAP of the random model.

$$AR = \frac{a_R}{a_P}$$

It is easy to see that for real rating models, the AR ranges from zero to one, and th closer the AR is to one, the better the rating model.

2.3. ROC and the AUROC curve

The construction of a ROC curve is illustrated in Figure 8.2, which shows possible dis tributions of rating scores for defaulting and non-defaulting debtors. For a perfect ratin model, the left distribution and the right distribution in Figure 8.2 would be separate For real rating systems, perfect discrimination in general is not possible. Distribution will overlap as illustrated in Figure 8.2 [reproduced from BCBS Working paper No.1 (2005)].

Assume one has to use the rating scores to decide which debtors will survive durin the next period and which debtors will default. One possibility for the decision-make would be to introduce a cut-off value C as in Figure 8.2, then each debtor with a ratin score lower than C is classed as a potential defaulter, and each debtor with a rating scor higher than C is classed as a non-defaulter. Four decision results would be possible. I the rating score is below the cut-off value C and the debtor subsequently defaults, th decision was correct. Otherwise, the decision-maker wrongly classified a non-defaulte as a defaulter. If the rating score is above the cut-off value and the debtor does no default, the classification was correct. Otherwise, a defaulter was incorrectly assigned t the non-defaulters' group.

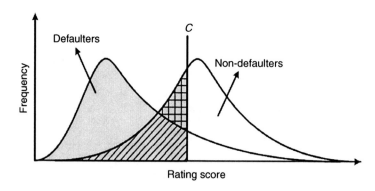

Figure 8.2 Distribution of rating scores for defaulting and non-defaulting debtors

Then, one can define a hit rate $HR(C)$ as

$$HR(C) = \frac{H(C)}{N_D},$$

where $H(C)$ is the number of defaulters predicted correctly with the cut-off value C and N_D is the total number of defaulters in the sample. This means that the hit rate is the fraction of defaulters who were classified correctly for a given cut-off value C. The false alarm rate $FAR(C)$ is then defined as

$$FAR(C) = \frac{F(C)}{N_{ND}},$$

where $F(C)$ is the number of false alarms, i.e. the number of non-defaulters who were classified incorrectly as defaulters by using the cut-off value C. The total number of non-defaulters in the sample is denoted by N_{ND}. In Figure 8.2, $HR(C)$ is the area to the left of the cut-off value C under the score distribution of the defaulters (coloured plus hatched area), whereas $FAR(C)$ is the area to the left of C under the score distribution of the non-defaulters (chequered plus hatched area).

To construct the ROC curve, the quantities $HR(C)$ and $FAR(C)$ are computed for all possible cut-off values of C that are contained in the range of the rating scores; the ROC curve is a plot of $HR(C)$ versus $FAR(C)$, illustrated in Figure 8.3.

The accuracy of a rating model's performance increases the steeper the ROC curve is at the left end and the closer the ROC curve's position is to the point $(0,1)$. Similarly, the larger the AUROC curve, the better the model. This area is called AUROC and is denoted by A. By means of a change of variable, it can be calculated as

$$A = \int_0^1 HR(FAR)d(FAR)$$

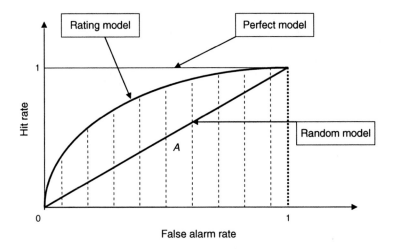

Figure 8.3 Receiver operating characteristic curves

The area A is 0.5 for a random model without discriminative power, and it is 1.0 for a perfect model. In practice, it is between 0.5 and 1.0 for any reasonable rating model. It has been shown in Engelmann *et al.* (2003) that

$$AR = \frac{a_R}{a_P} = \frac{N_{ND}(A - 0.5)}{0.5\,N_{ND}} = 2(A - 0.5) = 2A - 1.$$

2.4. Some further statistical properties of ROC measures

The ROC stems from the ODG. Assume we have two sets of continuous random variables X and Y. Let C be an arbitrary constant. Define

$$y = \text{Prob}\,(Y \le C) = F_Y(C) \quad \text{and} \quad x = \text{Prob}\,(X \le C) = F_X(C),$$

where x and y lie between $[0, 1]$ and C lies in $(-\infty, +\infty)$. Then the ODG is simply a plot of y against x. See Figure 8.4. There are some straight forward properties of the ODG:

1. The ODG curve is never decreasing, as x increases y cannot decrease.
2. If $\text{Prob}\,(Y \le C) = \text{Prob}\,(X \le C)$, then x and y are identically distributed, $y = x$ and the ODG curve is a 45° line.
3. If X first-order stochastic dominates (FSD) Y, then the ODG curve lies above the 45° line and vice versa.

Proof

$$\text{If } X \text{ FSD } Y \Rightarrow F_X(C) \le F_Y(C), \forall C \in \mathbb{R} \Rightarrow x \le y$$
$$\Rightarrow (x, y) \text{ lies above the 45° line}$$

If we regard y as score signals of defaulters in the next predefined period and x as those of the non-defaulters, then we expect any sensible rating system to produce $\text{Prob}\,(Y \le C) \ge \text{Prob}\,(X \le C)$ for all C. Thus, $x \le y$ for all C, and the ODG curve is above the 45° line. In the risk literature, it is referred to as the ROC curve. In terms of Section 2.3, y is the $HR(C)$ and x is the $FAR(C)$.

By assuming the existence of PDFs of F_X and F_Y, i.e. that they are both absolutely continuous, the following lemma can be derived:

Lemma 1

$$\text{If } x = F_X(C), \quad C = F_X^{-1}(x), \quad \text{then } \frac{\partial C}{\partial x} = \frac{1}{f_X(C)}, \quad \text{where } f_X(C) \text{ is the PDF of } X.$$

Proof

$$1 = \frac{\partial F_X(C)}{\partial x} = \frac{\partial F_X(C)}{\partial C} \cdot \frac{\partial C}{\partial x} = f_X(C)\frac{\partial C}{\partial x}, \quad \therefore \frac{\partial C}{\partial x} = \frac{1}{f_X(C)}.$$

Lemma 2

If $y = F_Y(C)$, then $\dfrac{\partial y}{\partial x} = \dfrac{f_Y(C)}{f_X(C)}$.

We see from Lemma 2 that the slope of the ODG curve is just the LR of Y and X evaluated at C.

Proof

$$\frac{\partial y}{\partial x} = \frac{\partial F_Y(C)}{\partial x} = \frac{\partial F_Y(C)}{\partial C} \cdot \frac{\partial C}{\partial x} = f_Y(C) \cdot \frac{\partial C}{\partial x} = \frac{f_Y(C)}{f_X(C)}.$$

Theorem

If $f_Y(C)/f_X(C)$ is increasing in C, then $\partial y/\partial x$ is increasing and the ODG curve is convex.

If $f_Y(C)/f_X(C)$ is decreasing in C, then $\partial y/\partial x$ is decreasing and the ODG curve is concave.

The latter case is the one that we are interested in, as it is the ROC curve. See Figure 8.4.

We have assumed that X and Y have PDFs, thus they are continuous random variables. Then, AUROC can then be expressed as

$$A = \text{Prob}(y \le x) \int_{-\infty}^{\infty} \text{Prob}(Y \le X | X = C)\, \text{Prob}(X = C)\, dC.$$

As X and Y are scores from different obligor groups, they are independent.
We have $\text{Prob}(Y \le X | X = C) = \text{Prob}(Y \le C)$.
Since $y = F_Y(C) = \text{Prob}(Y \le C)$ and $\partial x = f_X(C)\,\partial C$

$$A = \int_{-\infty}^{\infty} F_Y(C) f_X(C)\, dC = \int_{0}^{1} F_Y\left(F_X^{-1}(x)\right) dx. \tag{8.1}$$

A modelling exercise may choose a rating model that maximizes the AUROC with respect to the obligor group under review. But how would one estimate $\text{Prob}(y \le x)$?

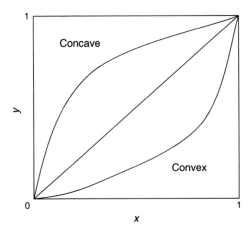

Figure 8.4 Ordinal dominance graph

Bamber (1975) and Engelmann *et al.* (2003) show that given a rating score sample of the obligors assigned by a rating model, the AUROC could be estimated in an non-parametric way using the Mann–Whitney U statistic.

On the other hand, if we had a parametric distribution model for X and Y, then we could explicitly derive a formula for the ROC curve and for the AUROC. In the next section, we will review some plausible distributions for X and Y and derive the closed-form solutions wherever possible.

3. Choices of distributions

In this section, we derive analytical formula for the ROC by assuming that the rating scores produced by rating models follow plausible distributions. The distributions we present here are Weibull distribution (including exponential distribution), logistic distribution, normal distribution and mixed models for X and Y. Where we have explicit closed forms for the ROC curve, we derive the closed form AUROC as well. The case of mixed distributions for X and Y can easily be extended from the following discussion.

We use the symbol M for the location parameters (sometimes the mean, sometimes the minimum), λ for the scale parameter and α for the shape.

3.1. Weibull distribution

We first present solutions under a Weibull distribution assumption of rating scores. The Weibull distribution is flexible and rich. A three-parameter Weibull distribution cumulative probability function (CDF) is given by

$$F(z) \equiv P(Z \le z) = 1 - e^{-\left(\frac{z-M}{\lambda}\right)^{\alpha}},$$

where $z > M$, $\alpha > 0$ and $\lambda > 0$. The inverse CDF of a three-parameter Weibull distribution is

$$F^{-1}(p) = M + \lambda\left[-\ln(1-p)\right]^{\frac{1}{\alpha}},$$

where $p \in [0, 1]$. Assuming y is the $HR(C)$ and x is the $FAR(C)$, the three-parameter Weibull distribution ROC is derived as

$$x = F_X(C) = 1 - e^{-\left(\frac{C-M_x}{\lambda_x}\right)^{\alpha_x}} \Rightarrow C = M_x + \lambda_x\left[-\ln(1-x)\right]^{\frac{1}{\alpha_x}}$$

$$y = F_Y(C) = P(Y \le C) = 1 - e^{-\left(\frac{C-M_y}{\lambda_y}\right)^{\alpha_y}}$$

$$\text{ROC}: y = 1 - \exp\left(-\left\{\frac{\lambda_x}{\lambda_y}\left[-\ln(1-x)\right]^{\frac{1}{\alpha_x}} + \left(M_x - M_y\right)\right\}^{\alpha_y}\right) \qquad (8.2)$$

The above ROC formula is very difficult to use for deducing an explicit closed-form formula for the AUROC although a numerical solution exists in this case once all the parameters are estimated. However, the situation becomes much better if we impose the slightly stronger assumption that the shape parameters α of the F_X and F_Y are equal to one.

We then have analytical closed-form solutions, and the Weibull distribution degenerates to a truncated exponential distribution family if M is positive and an extended exponential if M is negative.

Assume that $\alpha_x = \alpha_y = 1$, we then rewrite the above equation as follows:

$$ROC: y = 1 - e^{\frac{M_y - M_x}{\lambda_y}} (1 - x)^{\frac{\lambda_x}{\lambda_y}} .$$

Let $K = e^{\frac{M_y - M_x}{\lambda_y}}$ and $\theta = \frac{\lambda_x}{\lambda_y}$, then

$$AUROC = \int_0^1 \left[1 - K(1-x)^\theta\right] dx = 1 - \frac{K}{1+\theta} = 1 - \frac{\lambda_y}{\lambda_x + \lambda_y} e^{\frac{M_y - M_x}{\lambda_y}}, \qquad (8.3)$$

We next discuss some of the properties of Equation 3.

Property 1 for the two-parameter Weibull model

$M_y - M_x \in (-\infty, 0)$ for any plausible rating system. The smaller $M_y - M_x$ is (or the larger $|M_y - M_x|$ is), the closer the AUROC is to 1. Recall that M is the location parameter, in this case the minimum. Therefore, the rating system will receive a higher AUROC if it can better discriminate the defaulter from non-defaulter by the difference in their minimum values.

It is also interesting to see that if $M_y - M_x \to 0$, $AUROC \to 1 - \frac{\lambda_y}{\lambda_x + \lambda_y}$. As we illustrated in an earlier section, AUROC of a plausible, non-random rating system is above 0.5. This implies that the value of the scale parameters to which the rating scores are assigned have to be such that $0 < \lambda_y \leq \lambda_x$. Note that this condition is implied where both groups are exponential but also where both groups are truncated or extended exponentials with the same minima.

Property 2 for the two-parameter Weibull model

AUROC is monotonically increasing with respect to λ_x but monotonically decreasing with respect to λ_y.

Proof

$$\frac{d}{d\lambda_y} AUROC = \frac{K}{(\lambda_x + \lambda_y)^2 \lambda_y} \left[(M_y - M_x)(\lambda_x + \lambda_y) - \lambda_y(\lambda_x + 2\lambda_y)\right]$$

$$\left.\begin{array}{l} \left[(M_y - M_x)(\lambda_x + \lambda_y) - \lambda_y(\lambda_x + 2\lambda_y)\right] < (\lambda_x + \lambda_y)(M_y - M_x - \lambda_y) < 0 \\ \text{Since plausible rating systems, we expect } M_y \leq M_x, K \geq 0 \text{ and } \lambda_y > 0 \end{array}\right\} \Rightarrow$$

$$\Rightarrow \frac{d}{d\lambda_y} AUROC \leq 0$$

3.2. Logistic distribution

A two-parameter logistic distribution CDF is

$$F(z) \equiv P(Z \leq z) = \frac{e^{\frac{z-M}{\lambda}}}{1 + e^{\frac{z-M}{\lambda}}},$$

where $z \in \mathbb{R}$ and $\lambda > 0$. Here M is a mean parameter.

The inverse CDF of a two-parameter logistic distribution is

$$F^{-1}(p) = M + \lambda \ln\left(\frac{p}{1-p}\right),$$

where $p \in [0, 1]$. Again, assuming y is the $HR(C)$ and x is the $FAR(C)$, we have

$$x = F_X(C) = \frac{e^{\frac{C-M_x}{\lambda_x}}}{1 + e^{\frac{C-M_x}{\lambda_x}}} \Rightarrow C = M_x + \lambda_x \ln\left(\frac{x}{1-x}\right)$$

and

$$y = F_Y(C) = F_Y\left(F_X^{-1}(x)\right) = \frac{e^{\frac{C-M_y}{\lambda_y}}}{1 + e^{\frac{C-M_y}{\lambda_y}}} = \frac{e^{\frac{M_x-M_y}{\lambda_y}}\left(\frac{x}{1-x}\right)^{\frac{\lambda_x}{\lambda_y}}}{1 + e^{\frac{M_x-M_y}{\lambda_y}}\left(\frac{x}{1-x}\right)^{\frac{\lambda_x}{\lambda_y}}}.$$

Similarly to the Weibull distribution case, the AUROC with the above ROC specification can always be evaluated numerically. Moreover, by assuming $\lambda_x = \lambda_y = 1$, and in what follows assume that K does not equal 1, $K = e^{M_x - M_y}$, the above ROC equation can be simplified to

$$y = \frac{e^{M_x-M_y}x}{1 - x + e^{M_x-M_y}x} = \frac{Kx}{1 + (K-1)x}.$$

The AUROC can now be derived analytically.

$$AUROC = K\int_0^1 \frac{x}{1+(K-1)x}dx = \frac{K}{K-1}\int_0^1 \frac{(K-1)x}{1+(K-1)x}dx$$

Let $u = 1 + (K-1)x$,

$$AUROC = \frac{K}{(K-1)^2}\int_1^K \left(1 - \frac{1}{u}\right)du = \frac{K}{K-1}\left(1 - \frac{\ln K}{K-1}\right) \Rightarrow \lim_{K\to\infty} AUROC = 1$$

3.3. Normal distribution

A two-parameter normal distribution CDF is

$$F(z) \equiv P(Z \le z) = \int_{-\infty}^z \frac{1}{\lambda\sqrt{2\pi}}e^{\frac{1}{2}\left(\frac{x-M}{\lambda}\right)^2}dx = \Phi\left(\frac{z-M}{\lambda}\right),$$

where $z \in \mathbb{R}$ and $\Phi(.)$ is the standard normal probability distribution function. The inverse CDF of a two-parameter logistic distribution is

$$F^{-1}(p) = M + \lambda\Phi^{-1}(p),$$

where $p \in [0, 1]$. For y is the $HR(C)$ and x is the $FAR(C)$, we have

$$x = \Phi\left(\frac{C - M_x}{\lambda_x}\right) \Rightarrow C = M_x + \lambda_x \Phi^{-1}(x)$$

$$y = \Phi\left(\frac{C - M_y}{\lambda_y}\right) = \Phi\left(\frac{(M_x - M_y) + \lambda_x \Phi^{-1}(x)}{\lambda_y}\right),$$

which gives the ROC curve.

Therefore, $AUROC = \int_0^1 \Phi\left(\frac{(M_x - M_y) + \lambda_x \Phi^{-1}(x)}{\lambda_y}\right) dx$

This function can easily be evaluated numerically to obtain the AUROC.

Property
AUROC increases with $M_x - M_y$, and in particular if $M_x - M_y = 0$, $AUROC = 0.5$.
Proof and more properties can be found in Appendix.

3.4. Mixed models

It is obvious that as long as we have parametric distribution families for the defaulters and non-defaulters, we can always calculate an AUROC for the two-score samples from Equation 1 in Section 2, even with two different parametric distributions for the two populations.

4. Performance evaluation on the AUROC estimation with simulated data

Using simulated data, we carry out performance evaluations on AUROC estimations using both the non-parametric Mann–Whitney statistic and the analytic approach suggested in this chapter.

We first assume some known parametric distributions for the credit scores of defaulters and non-defaulters, and by doing this, we know the theoretical value of the AUROC. After generating simulated sample data from the assumed distributions for defaulter and non-defaulter scores, we estimate the AUROC and its CI using the two approaches. We repeat the simulation and estimation procedures a number of times. We then compare the accuracy of the AUROC estimation and the CI of the two approaches. Finally, we change the parameter values of the assumed distribution and repeat the simulation. We repeat the above procedures to evaluate the performance of the two approaches subject to different theoretical AUROC index values with different defaulter sample sizes. We choose the following distributions: two-parameter normal distributions, one-parameter exponential distributions and Weibull distributions with various shape and scale parameters.

4.1. Performance evaluations under the normal distribution assumption

We assume normal distributions for our parametric distribution of the credit scores of both defaulters and non-defaulters. The theoretical value of AUROC for the normal score samples is evaluated numerically.[1] The non-parametric estimate of AUROC is carried out using the ROC module in SPSS, and we use the bootstrap to re-sample 1000 replications to obtain the estimates of the analytic approach that also generates a two-sided 95% CI. The parameters of the parametric distribution are estimated for each replication and substituted back into the analytic AUROC formula. We then define the error as the difference between model estimates based on a sample and the theoretical AUROC value and compare the mean error and mean absolute error for the two approaches. The width of the CI is also compared.

We generate 50 normal samples from six different settings. Settings 1, 2 and 3, consisting of group 1, target the AUROC at a low value, whereas settings 4, 5 and 6, group 2, target the AUROC at a high value. Within each group there are three defaulter sample sizes: 20, 100 and 500. Credit rating models can be applied to at least three different types of groups: credit risk with corporate, counter party default risk in trading books, and credit risk in credit card and other loan type banking books. The default sample of corporate is usually small, such as 50 in 10 years, especially under a good economic cycle. Meanwhile, the number of defaults in a loan book or a credit card book in a commercial bank's banking book can be fairly large, usually in excess of several hundreds a year. The reason for selecting different defaulter sample sizes is to assess for which type of problem the analytic approach outperforms the non-parametric approach. We define a performance statistic as follows

Difference = non-parametric estimate − analytic estimate

Difference = non-parametric estimate − analytic estimate.

In Tables N1–N6, all mean CI widths show that the estimates of the analytic approach are better than the non-parametric estimates. As for the mean error and the mean absolute error, analytic estimates outperform the non-parametric estimates in Tables N1, N2 and N4–N6. Ratio to N shows the percentage difference from the non-parametric

Normal settings 1–3

| Sample | Normal distributions | | Number of observations | | |
	Mean	Standard deviation	Setting 1	Setting 2	Setting 3
X	2	2	1000	1000	1000
Y	1	3	20	100	500

Theoretical $AUROC$ = 0.609239.
Results on estimation error with 50 simulated normal samples and 1000 replication bootstrap.

Setting N1

Approach	Non-parametric	Analytic	Difference	Ratio to N
Mean error	0.007482	0.007466	0.000016	
Mean ABS error	0.050164	0.048408	0.001756	3.50%
Mean CI width	0.289225	0.276522	0.012703	4.39%

Setting N2

Approach	Non-parametric	Analytic	Difference	Ratio to N
Mean error	0.000404	−0.081959	0.082363	
Mean ABS error	0.025946	0.024957	0.000989	3.81%
Mean CI width	0.136364	0.130728	0.005636	4.13%

Setting N3

Approach	Non-parametric	Analytic	Difference	Ratio to N
Mean error	0.002643	0.002752	−0.000109	
Mean ABS error	0.014172	0.014809	−0.000636	−4.49%
Mean CI width	0.064965	0.062608	0.002357	3.63%

Normal settings 4–6

	Normal distributions		Number of observations		
Sample	Mean	Standard deviation	Setting 4	Setting 5	Setting 6
X	2	0.5	1000	1000	1000
Y	1	1	20	100	500

Theoretical $AUROC = 0.814448$.
Results on estimation error with 50 simulated normal samples and 1000 replication bootstrap.

Setting N4

Approach	Non-parametric	Analytic	Difference	Ratio to N
Mean error	0.009138	0.006718	0.002421	
Mean ABS error	0.046588	0.045725	0.000863	1.85%
Mean CI width	0.232187	0.215922	0.016265	7.01%

Setting N5

Approach	Non-parametric	Analytic	Difference	Ratio to N
Mean error	0.001187	0.000510	0.000678	
Mean ABS error	0.025564	0.024311	0.001253	4.90%
Mean CI width	0.112444	0.107148	0.005296	4.71%

Setting N6

Approach	Non-parametric	Analytic	Difference	Ratio to N
Mean error	0.001470	0.001303	0.000167	
Mean ABS error	0.012239	0.011061	0.001178	9.62%
Mean CI width	0.052653	0.049464	0.003189	6.06%

approach estimate. The larger the ratio to N, the more the analytic approach outperforms the non-parametric approach.

4.2.   Performance evaluations under exponential distribution assumption

In this performance evaluation, we assume exponential distributions for our parametric distribution of the credit scores of both the defaulters and the non-defaulters. The theoretical value of AUROC for the exponential score samples is evaluated analytically by the closed-form formula 8.3 in Section 3.1. The performance evaluation setting is very similar to that with normal distribution. Again there are six settings across different AUROC values and defaulter sample sizes.

In Tables E1–E6, all the mean absolute error and the mean CI widths show that the estimates of the analytic approach are better than the non-parametric estimates. Ratio

Exponential settings 1–3

	Exponential distributions	Number of observations		
Sample	Scale parameter (Lamda)	Setting 1	Setting 2	Setting 3
X	3	1000	1000	1000
Y	1.5	20	100	500

Theoretical $AUROC = 0.666667$.
Results on estimation error with 50 simulated normal samples and 1000 replication bootstrap.

Setting E1

Approach	Non-parametric	Analytic	Difference	Ratio to N
Mean error	−0.008179	−0.007035	−0.001144	
Mean ABS error	0.040993	0.040056	0.000938	2.29%
Mean CI width	0.209540	0.189586	0.019954	9.52%

Setting E2

Approach	Non-parametric	Analytic	Difference	Ratio to N
Mean error	−0.000987	0.000034	−0.001021	
Mean ABS error	0.025320	0.021922	0.003398	13.42%
Mean CI width	0.099043	0.088280	0.010763	10.87%

Setting E3

Approach	Non-parametric	Analytic	Difference	Ratio to N
Mean error	−0.002926	−0.003401	0.000475	
Mean ABS error	0.011471	0.011015	0.000456	3.98%
Mean CI width	0.055636	0.047672	0.007964	14.31%

Exponential settings 4–6

	Exponential distributions		Number of observations		
Sample	Scale parameter (Lambda)		Setting 4	Setting 5	Setting 6
X	4		1000	1000	1000
Y	1		20	100	500

Theoretical AUROC = 0.800000.
Results on estimation error with 50 simulated normal samples and 1000 replication bootstrap.

Setting E4

Approach	Non-parametric	Analytic	Difference	Ratio to N
Mean error	−0.008576	−0.006721	−0.001855	
Mean ABS error	0.033790	0.031174	0.002616	7.74%
Mean CI width	0.145500	0.132758	0.012742	8.76%

Setting E5

Approach	Non-parametric	Analytic	Difference	Ratio to N
Mean error	0.002783	0.003403	−0.000621	
Mean ABS error	0.015655	0.014320	0.001335	8.53%
Mean CI width	0.071140	0.064132	0.007008	9.85%

Setting E6

Approach	Non-parametric	Analytic	Difference	Ratio to N
Mean error	0.000118	0.000521	−0.000403	
Mean ABS error	0.007710	0.007495	0.000215	2.79%
Mean CI width	0.043742	0.034280	0.009462	21.63%

to N shows that the non-parametric approach estimates provide a significantly better CI than the non-parametric estimates.

4.3. Performance evaluations under Weibull distribution assumption

In this performance evaluation, we assume Weibull distributions with scale and shape parameters for our parametric distribution of the credit scores of both the defaulters and the non-defaulters. The theoretical value of AUROC for the Weibull score samples is evaluated analytically by the closed-form formula 8.2 in Section 3.1 by setting the location parameters to zero. The maximum estimation of sample distribution parameters is obtained by a numerical approximation. As we have a shape parameter for the Weibull distribution that may shift the shape of the distribution significantly, we evaluate the performance of the two approaches under two cases: with the same shape parameter for defaulter and non-defaulter sample and with different shape parameters. The theoretical value of AUROC for the normal score samples is also evaluated numerically (Note 1). The rest of the performance evaluation setting is very similar to that of the normal distribution. Here also, there are six settings across different AUROC values and defaulter sample sizes.

In Tables W1–W6, all mean CI widths show that the estimates of the analytic approach are marginally better than the non-parametric estimates. As for the mean error and the mean absolute error, analytic estimates marginally outperform the non-parametric estimates in Tables W2, W3, W5 and W6. Because we use numerical approximation for sample maximum likelihood estimates and because the estimation error could be

Weibull settings 1–3

| Sample | Weibull distributions | | Number of observations | | |
	Shape parameter	Scale	Setting 1	Setting 2	Setting 3
X	2	2	1000	1000	1000
Y	1	1	20	100	500

Theoretical AUROC = 0.757867.
Results on estimation error with 50 simulated normal samples and 1000 replication bootstrap.

Setting W1

Approach	Non-parametric	Analytic	Difference	Ratio to N
Mean error	0.005128	0.010230	−0.005102	
Mean ABS error	0.051701	0.054179	−0.002478	−4.79%
Mean CI width	0.242836	0.226842	0.015994	6.59%

Setting W2

Approach	Non-parametric	Analytic	Difference	Ratio to N
Mean error	0.001110	0.000983	0.000127	
Mean ABS error	0.022661	0.022363	0.000298	1.32%
Mean CI width	0.112448	0.109910	0.002538	2.26%

Setting W3

Approach	Non-parametric	Analytic	Difference	Ratio to N
Mean error	0.0027541	0.0030963	−0.000342	
Mean ABS error	0.0123445	0.0118544	0.000490	3.97%
Mean CI width	0.0548159	0.0533400	0.001476	2.69%

Weibull settings 4–6

Weibull distributions			Number of observations		
Sample	Shape parameter	Scale	Setting 4	Setting 5	Setting 6
X	1	3	1000	1000	1000
Y	1	1	20	100	500

Theoretical AUROC = 0.75.
Results on estimation error with 50 simulated normal samples and 1000
replication bootstrap.

Setting W4

Approach	Non-parametric	Analytic	Difference	Ratio to N
Mean error	0.000084	0.000314	−0.000231	
Mean ABS error	0.035960	0.036155	−0.000195	−0.54%
Mean CI width	0.168248	0.165242	0.003006	1.79%

Setting W5

Approach	Non-parametric	Analytic	Difference	Ratio to N
Mean error	0.003680	0.003795	−0.000115	
Mean ABS error	0.018331	0.017988	0.000343	1.87%
Mean CI width	0.082652	0.081830	0.000822	0.99%

Setting W6

Approach	Non-parametric	Analytic	Difference	Ratio to N
Mean error	0.003889	0.003961	−0.000072	
Mean ABS error	0.009632	0.009525	0.005340	1.11%
Mean CI width	0.048446	0.047586	0.000860	1.77%

fairly large when we have a small sample, we observe that this estimation error is passed through our analytic estimation for the AUROC index making the mean absolute errors estimated from the analytic approach larger than the non-parametric approach in settings W1 and W4. This also reduces the gain of the analytic approach over the non-parametric approach when compared with the previous tests.

5. Summary

Although the analytic approach gives no better estimates than the non-parametric one when we use approximated maximum likelihood estimates for small samples, the performance evaluation shows that the analytic approach works at least as well as the non-parametric approach in the above tests and, in most cases, provides better mean absolute error estimates and CI estimates.

The above discussion has the following implications. If appropriate parametric distributions for the defaulter and non-defaulter scores can be identified, then the AUROC and its CI can be estimated more accurately using the analytic approach. On the other hand, if the rating model can be designed so that the score sample is generated by some specific parametric distribution families, then a better rating model could be found by using the analytic AUROC as the objective function to maximize in the model selecting process.

Another interesting finding is the effect of defaulter sample size on AUROC. The above experiments clearly show the level of estimation error in both methods with different sample sizes, and the error can be substantially large only if we have a small defaulter sample.

In addition, although it is not very clear from the results in Sections 4.1 and 4.2, the analytic approach seems to provide more gain over the non-parametric approach when the AUROC index is in its high-value region than in its low-value region. The reason for this is not clear, so more research is needed.

6. Conclusions

This chapter reviews some of the prevailing credit rating model validation approaches and, in particular, studies the analytic properties of the ROC curve and its summary index AUROC. We use the concept of the population ROC curve to build analytic models of ROC curves. It has been shown through simulation studies that greater accuracy of estimation relative to the non-parametric methods can be achieved. We also show that there are some situations where the accuracy gain of the analytic ROC model may decrease, a finding that should be taken into account when applying the analytic models to practical applications.

Moreover, with some distributions, where the closed-form solution of AUROC is available, analytic AUROC can be directly used as an objective function to maximize during the rating model selection procedure. This means that if the rating scores can be transformed into those distributions, analytic AUROC could offer a powerful model selection tool.

Finally, we also studied the performance of both non-parametric and analytic ROC models under different defaulter sample size, research that had not been done previously. The error size can be substantially significant when we have a small defaulter sample, a frequently met situation in corporate credit risk studies and in periods of benign macro-economic background.

7. Acknowledgements

Wei Xia thanks Birkbeck College for the generous funding support and Ron Smith for his helpful comments.

References

Bamber, D. (1975) The area above the ordinal dominance graph and the area below the receiver operating characteristic graph. *Journal of Mathematical Psychology*, 12, 387–415.

Basel Committee on Banking Supervision (BCBS) (2004) *International Convergence of Capital Measurement and Capital Standards: A Revised Framework*. Bank for International Settlements, June.

Basel Committee on Banking Supervision (BCBS) (2005) Studies on the Validation of Internal Rating Systems, Working paper No. 14.

Engelmann, B., Hayden, E. and Tasche, D. (2003), Testing rating accuracy. *Risk*, January, 16, 82–6.

Sobehart, J.R. and Keenan, S.C. (2001) Measuring default accurately. *Risk Magazine*, March, 14, 31–3.

Sobehart, J.R. and Keenan, S.C. (2004) Performance evaluation for credit spread and default risk models. In *Credit Risk: Models and Management* (David Shimko, ed.), second edition. London: Risk Books, pp. 275–305.

Sobehart, J.R., Keenan, S.C. and Stein, R. (2000) Validation. methodologies for default risk models. *Credit*, 1, 51–6.

Swets, J.A. (1988) Measuring the accuracy of diagnostic systems. *Science*, 240, 1285–93.

Note

1. The theoretical AUROC is approximated by 100 000 partitions, whereas the bootstrap estimation is approximated by 10 000 partitions.

Appendix

1. The properties of AUROC for normally distributed sample

Property 1
AUROC increases with $M_x - M_y$, and in particular, if $M_x - M_y = 0$, AUROC $= 0.5$.

Proof   For inverse normal distribution function $u = \Phi^{-1}(v)$, $v \in [0, 1]$ and $u \in (-\infty, +\infty)$. It is an odd function in orthogonal coordinates with centre of $(v = 0.5, u = 0)$.

For cumulative normal distribution function, $t = \Phi(u)$. This is also an odd function in orthogonal coordinates with centre of $(u = 0, t = 0.5)$.

It follows that $f(x) = \Phi\left[(\lambda_x/\lambda_y)\,\Phi^{-1}(x)\right]$ is also an odd function in orthogonal coordinates with centre of $(x = 0.5, f(x) = 0.5)$, when $M_x - M_y = 0$. Rewrite $f(x)$ as follows:

$$f(x) = [f(x) - 0.5] + 0.5 = g(x) + 0.5,$$

where $g(x)$ is an odd function with centre of $(x = 0.5, g(x) = 0)$. Then we can show that

$$\text{AUROC} = \int_0^1 f(x)dx = \int_0^1 g(x)dx + \int_0^1 0.5dx = \int_0^{0.5} g(x)dx + \int_{0.5}^1 g(x)dx + \int_0^1 0.5dx$$

$$= \int_0^{0.5} g(x)dx - \int_0^{0.5} g(x)dx + \int_0^1 0.5dx = \int_0^1 0.5dx = 0.5 \quad \text{QED.}$$

The above property is also quite intuitive. If the means of two normally distributed populations equal each other, then overall there is no discriminatory power of the models based on this rating mechanism, i.e. neither X or Y FSD. So the AUROC is 0.5. A special case for this is when we have two identical distributions for X and Y. Then second-order stochastic dominance (SSD) cannot be identified by AUROC when $M_x - M_y = 0$.

Property 2
The relations with λ_x and λ_y are slightly more complicated.

$$AUROC \begin{cases} \in (0.5, 1), & \text{decreases with } \lambda_x, & \text{when} & M_x - M_y > 0 \\ = 0.5, & \text{irrelavent to } \lambda_x, & \text{when} & M_x - M_y = 0 \\ \in (0, 0.5), & \text{increases with } \lambda_x, & \text{when} & M_x - M_y < 0 \end{cases}$$

$$AUROC \begin{cases} \in (0.5, 1), & \text{decreases with } \lambda_y, & \text{when} & M_x - M_y > 0 \\ = 0.5, & \text{irrelavent to } \lambda_y, & \text{when} & M_x - M_y = 0 \\ \in (0, 0.5), & \text{increases with } \lambda_y, & \text{when} & M_x - M_y < 0 \end{cases}$$

We are only interested in the rating models, and this is the case where X should FSD Y, i.e. $M_x - M_y > 0$, so it is clear that with smaller standards of the two normal distributions, the two samples are more separated than those with larger standard deviations when $M_x - M_y > 0$.

Figures 8.5 and 8.6 shows the AUROC with different Lambda settings. Lambda of X is written as L.X and Lambda of Y is L.Y.

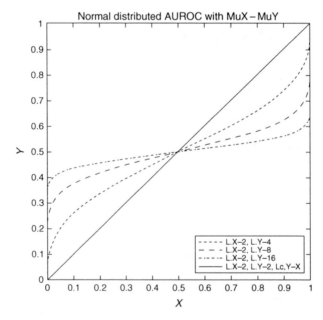

Figure 8.5 Normal distributed area under the receiver operating characteristic (AUROC) curve with the same mean

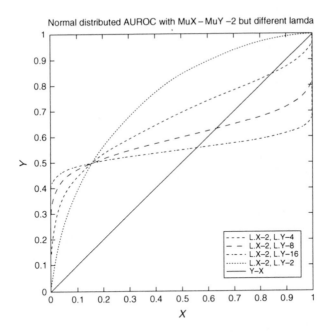

Figure 8.6 Normal distributed area under the receiver operating characteristic (AUROC) curve with different means

Remark

The closer the AUROC of a rating system is to 0.5, the less discriminatory power it has. The closer the AUROC of a rating system is to 0 or 1, the better its discriminatory power. Therefore, under the normally distributed scoring variable assumption, the smaller the variance, the better the discriminatory power the rating system has.

When $M_x - M_y < 0$, a scoring system would give defaulters, Y, higher scores. Hence, even the discriminatory power is higher when we have smaller variances on X and Y in this case, but the AUROC will be smaller.

9 The validation of equity portfolio risk models

*Stephen Satchell**

Abstract

The purpose of this chapter is to survey risk validation issues for equity portfolio models. Because risk is measured in terms of volatility, an unobservable time-varying variable, there are a number of difficulties that need to be addressed. This is in contrast to a credit risk model, where default or downgrading is observable. In the past, equity risk models have been validated rather informally, but the advent of Basel II has brought about a need for a more formal structure. Furthermore, a number of past high-profile court cases have considered the worthiness of the risk models used by fund managers as part of the broader question of whether the manager has been competently managing risk. To be able to demonstrate that the model user has been managing risk properly, it will be necessary, going forward, to demonstrate that back-testing and other forms of model validation have been practised; and that the systems in place have been designed for the efficient execution of such practise. There have been a number of insightful articles on portfolio risk written by the Faculty and Institute of Actuaries, see Gardner *et al.* (2000) and Brooks *et al.* (2002) among others. These papers address in some detail many of the problems concerning portfolio risk and make a number of recommendations. Any equity portfolio risk manager should read them. However, they do not say a great deal about risk model validation. Indeed, there are only occasional sections. For example, in Gardner *et al.* (2000), the following is stated in Section 2.76:

> 'It is claimed that risk models are subject to exhaustive tests by their owners, although the results are not routinely published.'

The authors point out that neither model user nor plan sponsor will find the level of internal validation sufficiently transparent to feel particularly comfortable. As in most Basel II contexts, it is clear that there is a need for external validation, and this chapter puts forward some preliminary suggestions and guidelines. This topic will clearly change with time, and there is no claim that the contents of this chapter will be best practised in a decade. In many papers written as a result of Basel I, much is made of back-testing. Without denigrating these efforts, we understand back-testing to mean a comparison of what the model sets out to do, and what actually happens. In this particular context, it simply means comparing model forecasts of portfolio risk with the risk that the portfolio actually experienced. Another issue that we deal with in a rather cursory manner is the metric used for measuring portfolio risk. Whilst downside risk measures have tremendous appeal at the univariate level, they are generically inappropriate for portfolio analysis, as there is simply a dearth of theorems linking the downside measure of the aggregate to the downside measures of the constituents. Downside

* Faculty of Economics, University of Cambridge and Trinity College, Cambridge, UK

enthusiasts regularly produce such results, but there is always a subtext. This subtext stems from the assumption of normality or ellipticity, in which case the downside risk metric is invariably a transformation of the standard deviation of absolute or relative returns. For the reasons stated in the above paragraph, portfolio risk in the context of this chapter means standard deviation or its square, variance. The whole analysis can be reworked as tracking error (TE), or TE squared, if we use benchmark relative returns. We give formal definitions in the text. We shall first discuss linear factor models (LFMs) and how they are used for forecasting risk. This may be slightly repetitive in terms of the material in other chapters, but it is convenient to keep this chapter self-contained, and in any case the emphasis here is different. We need to consider how the risk model forecasts can be evaluated. Thus, we will need to look closer at volatility as a temporal and cross-sectional process. We shall further discuss some of the non-mathematical issues associated with equity risk model validation. In Section 1, we cover LFMs, using the standard taxonomy of time series, statistical and fundamental. Sections 2–4 deal with each of these approaches in turn. Section 5 deals with forecast construction and evaluation, the latter in rather broad terms. Section 6 discusses diagnostics in some detail. Section 7 discusses issues to do with time horizon of forecast and frequency of model construction and data. Section 8 focuses on model residuals. Section 9 deals with Monte Carlo procedures and Section 10 concludes.

1. Linear factor models

These are fundamental to the evaluation of portfolio risk. They are the basis of equity risk modelling, and all major systems are based on them. Before we discuss risk modelling, we need to understand how linear factor models (LFMs) are constructed.

Suppose that $R_t \underset{(N \times 1)}{} =$ vector of N equity monthly returns possibly relative to cash. We now suppose that these returns are linearly related to K factors, whose values at time t are represented by the vector f_t

$$R_t = \alpha + \beta f_t + V_t \qquad (9.1)$$

In the simplest form α, β are fixed, f_t is $K \times 1$, $V_t \sim (0, D)$, meaning mean zero, covariance matrix D where D is diagonal. Diagonality implies that the factors are uncorrelated with each other; the i-th diagonal element corresponds to the idiosyncratic variance of the i-th stock return. It represents risk attributable to firm-specific events, not covered by the K common factors. It could be risk because of local effects, particular staff issues, an insane finance director etc.

Turning to the factors, we assume them to be uncorrelated with the errors, $\mathrm{Cov}(f_t, V_t') = 0$, $\mathrm{Cov}(f_t) = \Omega_{FF}$. The notation $\mathrm{Cov}(X)$, where X is a vector means 'the covariance matrix of X'; the notation $\mathrm{Cov}(X, Y)$, where X and Y are vectors will be a matrix of covariances between the elements of X and the elements of Y. Using standard results in multivariate statistics, we can get an expression for the covariance matrix of equity returns in terms of the constituent parts.

$$\Omega_{RR} = \beta \Omega_{FF} \beta' + D \qquad (9.2)$$

For a portfolio p with vector of weights, (w),

$$\sigma_P^2 = w' \Omega_{RR} w = w' \beta \Omega_{FF} \beta' w + w' D w = \sigma_s^2 + \sigma_R^2 \qquad (9.3)$$

There are three fundamental methods, time series versus statistical versus cross-sectional (CS) regression.

2. Building a time series model

For time series models, the assumption is that we have observable factors, and we estimate the parameters in the LFM. Typically, you might choose your factors in the context of a particular market you wish to model, say for the UK market, you might have the FT. All share return as a market factor and 10 or 38 sector variables as additional factors depending upon how fine you go down into factor detail. Taking the 10 sector case, and the market, as an 11 factor model, you can compute returns to all these factors directly by considering capitalization – weighted sector returns and estimating stock betas (exposures) – by conventional regression methods.

Thus

$$r_{it} = \alpha_i + \beta_i' f_t + \varepsilon_{it}; \quad \varepsilon_{it} \sim (0, \sigma_i^2), \quad t = 1, T; i = 1, N \qquad (9.4)$$

Note that this is in econometric terms a generalized least squares (GLS) problem as each equation has a different idiosyncratic variance, but as the model has common regressions, it is a theorem that system-wide GLS equals system-wide ordinary least squares (OLS). Now, it follows that optimal estimation is equivalent to OLS equation by equation. This assumes we impose no restrictions on parameters before estimation. Furthermore, if we assume joint normality, these estimators are maximum likelihood. In practise, there is additional information we may wish to include. In our example, it could be sector membership, so we may wish to make our estimation method sensitive to the fact that stock i is classified as a member of sector j. Once we impose restrictions on the equations, we need to consider some version of GLS again.

In system terms, after estimation, we have, where a tilde means an OLS estimate,

$$\underset{N\times 1}{\tilde{r}_t} = \underset{N\times 1}{\hat{\alpha}} + \underset{N\times k, k\times 1}{\hat{\beta} f_t} + \underset{N\times 1}{\hat{\varepsilon}_t} \qquad (9.5)$$

and we can estimate σ_i^2 by $\hat{\sigma}_i^2 = \sum_{t=1}^{T} \hat{\varepsilon}_{it}^2 / T$.

Notice that the residuals will have a sample mean of zero by construction as all equations are estimated with a constant. For any fixed weight portfolio P, $w(N \times 1)$

$$r_{pt} = \hat{\alpha}_p + \hat{\beta}_p' f_t + \hat{\varepsilon}_{pt}$$

$$\hat{\alpha}_p = w' \hat{\alpha}; \quad \hat{\beta}_p' = w' \hat{\beta}'; \quad \hat{\varepsilon}_{pt} = w' \hat{\varepsilon}$$

We need the notation $\text{Var}(X)$ to mean variance of X and $\text{Cov}(X, Y)$ to mean the covariance of X and Y where X and Y are now scalars. As in the capital asset pricing model (CAPM), the portfolio variance obeys the following:

$$\text{Var}(r_{pt}) = \text{Var}(\hat{\beta}'_p f_t) + \text{Var}(\hat{\varepsilon}_{pt})$$

The risk decomposition in this model between systematic and residual risk is exact, as by OLS properties $\text{Cov}(f_t, \hat{\varepsilon}_{pt}) = 0$, as long as the regression has been estimated with a constant. This procedure allows us a further ex post reality check, as we can evaluate factor risk versus residual risk that will equal total risk. Often, experienced modellers have strong priors as to what constitutes a sensible risk decomposition and implausible numbers here indicate a problem with the model.

3. Building a statistical factor model

Portfolio risk was discussed earlier, in Section 2, for the case when the factors were known. A time-series model was used and exposures (Bs) and residual variances (D) were estimated.

However, we may not wish to proceed this way. We may believe that our factors are unknown and our exposures are unknown but wish to estimate them.

In this case, as before

$$R_t = \alpha + \beta f_t + \varepsilon_t; \; \varepsilon_t \overset{iid}{\sim} (0, D); \; f_t \overset{iid}{\sim} (\mu_F, \Omega_F) \tag{9.6}$$

So that $E(R_t) = \alpha + \beta \mu_F$ and $\tag{9.7}$

$\text{Cov}(R_t) = \beta \Omega_F \beta' + D$ as before $\tag{9.8}$

The difference here is that we do not observe α, β, f_t or ε_t, just R_t. In the previous case we observed R_t and f_t. We shall discuss two methods: principal components and factor analysis. Principal components ignore all the structure in Equations 9.6, 9.7 and 9.8, whilst factor analysis tries to take it into account. If f_t is observable we are back to our earlier case. In general, it is not so to proceed we need to revise Eigen-vectors and Eigen-values and positive definite matrices. Our treatment follows the text of Dhrymes (1970) very closely.

3.1. Definitions

Here, we define some of the terms necessary to explain how principal components are constructed. In what follows, $\underset{(n \times n)}{\Omega}$ is a $(n \times n)$ covariance matrix. As will be clear, covariance matrices have an important property called positive definiteness, which we define next.

$\underset{(n \times n)}{\Omega}$ is positive definite if $w'\Omega w > 0, \forall w \neq 0$. If Ω is symmetric, P orthogonal, P such that $P\Omega P' = D_1$, where D_1 is diagonal, choosing $\underset{1 \times n}{w'} = (1, 0, \ldots, 0)$, we have $w'P\Omega P'$

$w = wD_1w = du$, the first diagonal element of $D > 0$. Furthermore, $w'P$ is the first row of P which is the first column of P', this column we denote by P_{01}.

As $P\Omega P' = D_1$, $P'P\Omega P' = P'D_1$, or $\Omega P' = P'D_1$, considering the first column of P', P_{01}, we have $\Omega P_{01} = P_{01}du$.

So, we see that du is the first Eigen-value of Ω and it is positive. Generally for a $(n \times n)$ matrix A, an Eigen-vector x is a $(n \times 1)$ vector such that $Ax = \lambda x$ where λ is a constant, the constant λ is the Eigen-value associated with the Eigen-vector x. An arbitrary $(n \times n)$ matrix will have n Eigen-vectors and n Eigen-values, although these are not necessarily distinct. That there are n of them is an interesting consequence of the fundamental theorem of algebra.

3.2. Principal components

We next discuss how to construct our new statistical variables/factors. Let $x \sim N(\mu, \Sigma)$, x being an m-dimensional vector, and let it be desired to find a linear combination of the elements of x having maximal variance.

If $\zeta' = \alpha'x$ is such a linear combination, then $\text{Var}(\zeta) = \alpha'\Sigma\alpha$

And it becomes clear that the problem does not have a solution, for if we choose $\alpha^* = c\alpha$.

For a fixed α, we can make the variance of ζ arbitrarily large by taking c to be large.

Hence, we have a scale problem; in this context, the problem is usually resolved by the convention $\alpha'\alpha = 1$, which requires that the vector defining the linear combination be of unit length. Notice that this is a rather poor normalization for financial data, as the linear combination we are considering will usually be a portfolio and hence will also need to be rescaled so that the weights add up to 1. This double normalization is rather a nuisance and can lead in practise to rather peculiar factors.

Thus, the problem becomes equivalent to the maximization of the Lagrangean

$$L = \alpha'\Sigma\alpha + \lambda(1 - \alpha'\alpha)$$

This yields the first-order conditions

$$\frac{\partial L}{\partial \alpha} = 2\Sigma\alpha - 2\lambda\alpha = 0$$

$$\frac{\partial L}{\partial \lambda} = 1 - \alpha'\alpha = 0$$

(9.9)

and it is verified that the solution vector obeys

$$\alpha'\Sigma\alpha = \lambda \tag{9.10}$$

On the contrary, the first set of equations in Equation 9.9 implies that λ must be one of the characteristic roots of Σ. As a result of Equation 9.10 and the nature of the problem, we must choose the largest characteristic root, say λ_1, and the (normalized) characteristic vector associated with it, say $\alpha_{(1)}$. Thus, we define the (first) principal component of (the elements of) x by $\zeta_1 = \alpha'_{(1)}x$. Where $\alpha_{(1)}$ obeys $(\lambda_1 I - \Sigma)\alpha_{(1)} = 0$ and $\alpha'_{(1)}\alpha_{(1)} = 1$ and λ_1 is the largest root of $|\lambda I - \Sigma| = 0$.

It follows that $E(\zeta_1) = \alpha'_{(1)}\mu$ and $\mathrm{Var}(\zeta_1) = \lambda_1$. To construct the second principal component, we repeat the calculations above but impose an orthogonality condition requiring that the second factor be uncorrelated with the first. Repeating this procedure, we end up with n linear combinations of equity returns, all uncorrelated with each other. These can be renormalized as portfolios and can become the factors in our LFM. They have certain advantages in that they have minimal data requirements; they have certain clear disadvantages in that they cannot be easily identified, and that the portfolios rapidly become ridiculous once we get past the first two or three factors.

We omit discussion of factor analysis. There are clear expositions of this in many texts; see, for example, Campbell *et al.* (1997). The critical difference is that the factors here are built by consideration of the risk decomposition described in Equation 9.2, whilst principal components work with the full covariance matrix and do not take into account the nature of the matrix structure. Whilst the former is more appealing, intellectually, it comes at a cost, as we need to impose a number of rather arbitrary identifiability conditions before we can identify the factors. Whilst we believe there are contexts where these approaches can be helpful, we do not think that they work particularly well for a conventional actively managed equity portfolio, where rebalancing is a relatively low-frequency event.

4. Building models with known beta's

Again, we have Equation 9.1, but we treat the exposures (β) and R_t as known, β is now β_t and we estimate f_t and D_t so our model is actually $R_t = \alpha_t i_N + \beta_t f_t + V_t; V_t \sim (0, D_t)$ and $\Omega_t = \beta_t \Omega_{F_t} \beta'_t + D_t$.

This is much more complex, and it is only possible to estimate it if we treat β_t and R_t as known.

These models are based on known betas; which are themselves attributes of the company. They would be things such as firm size, dividend yield and book to price ratio plus industry. These models are most frequently used and occupy at least 80% of the risk model market for portfolio risk. Because they use company accounts data to compute exposures, they are called fundamental models. As the betas, as defined, cover all stocks at each period of time, they are denoted by $B_{itj}; i = 1, N; t = 1, T; j = 1, k$. Our model now is

$$R_{it} = \alpha_t + \sum_{j=1}^{k} \beta_{itj} f_{jt} + V_{it}$$
$$\underset{N\times1}{V_t} \sim (0, D_t), \underset{k\times1}{f_t} \sim (\mu_{F_t}, \Omega_{FF_t}) \tag{9.11}$$

These models are compared with macroeconomic (time series models) and statistical (principal components, factor analysis) models by Connor (1995). In this paper, he argues in favour of fundamental models; however, other research could be quoted making the case for the other two.

In Equation 9.11, R_{it} and β_{itj} are known and we need to estimate $\{f_{jt}\}$; the implied factor returns. This is done by CS regressions.

Ideally, one should use GLS based on some rolling estimator of D (if we assume $D_t = D, \forall t$), i.e. if

$$D = \begin{pmatrix} \sigma_1^2 & 0 & \cdots \\ 0 & \ddots & \vdots \\ \vdots & \cdots & \sigma_N^2 \end{pmatrix}$$

and we estimate σ_i^2 by s_{it}^2, then run the CS regression.

$$\frac{R_{it}}{S_{it}} = \frac{\alpha_t}{S_{it}} + \sum \frac{B_{itj}f_{jt}}{S_{it}} + \frac{V_{it}}{S_{it}}$$

The transformed error $\underset{\sim}{\varepsilon}_t = \varepsilon_{it} = \frac{V_{it}}{s_{it}}$ will be approximately $(0, I_N)$.

Notice that if add a time-specific constant α_t, then, averaging over i, $\underset{i}{E}(V_{it}) = 0$. However, this does not mean that $\underset{t}{E}(V_{it}) = 0$ and, indeed, it may not. This becomes a problem if we wish to estimate σ_i^2 based on past time-series of (V_{it}); $t = 1, \ldots, T$. We may have to subtract residual means. These models can be used to construct a history of estimated $\{f'_{ts}\}$, which are then used to estimate Ω_{FF}. Together with our estimate of D, we can put all this together to form

$$\hat{\Omega}_{RR} = \beta \hat{\Omega}_{FF} \beta' + \hat{D} \tag{9.12}$$

Up to this point, all discussion has been in terms of absolute risk; in practise, many risk calculations are in terms of relative risk. This is referred to as TE. We need benchmark weights denoted by the vector b. The formal definition of TE is the standard deviation of relative returns measured annually; typical numbers will be 1%, 2% or 3%. From a risk management perspective, it is necessary to be absolutely clear about this definition. So, as a standard deviation, it will implicitly be subtracting the portfolio expected returns, even though the risk models may not explicitly model such quantities. There is a persistent confusion about TE in the industry, as many practitioners still think of it as the second central moment about zero. Now, it is clear that these two definitions will coincide if the mean of the benchmark equals the mean of the active fund. If the risk validator believes that the risk model is producing the second central moment, he or she will need to bring into play estimates of relative returns to make *ex ante* and *ex post* comparisons. The history of this confusion is quite interesting. The earliest application of TE was for index funds, where, in the early days, full replication was very challenging. Thus, the index tracker was trying to replicate the index with a limited number of stocks. In this context, the relative return was set up to be zero, and so the two concepts of TE coincided. Passing over some of these subtleties, we note that the portfolio analogue of Equation 9.12, measured in TE terms, can be expressed as

$$(w - b)' \hat{\Omega}_{RR}(w - b) = (w - b)' \beta \hat{\Omega}_{FF} \beta'(w - b) + (w - b)' \hat{D}(w - b) \tag{9.13}$$

Notice that this gives an explanation of

$$\text{Total relative risk squared} = (TE)^2 = \text{factor risk} + \text{idiosyncratic risk} \tag{9.14}$$

5.   Forecast construction and evaluation

All the above models described in Sections 2–4 can be used to forecast TE. To use them
you input w and b and the model gives you TE together with risk decompositions and
risk contribution, i.e. what factors induce what quantities of risk taken as a percentage
or measured in absolute terms?

All the models will take data up to the current period and estimate the equity covariance
matrix. Indeed, different risk models can be interpreted as different forecasts of the same
covariance matrix. Traditionally, these models have been reestimated each month, and
each month a new forecast is obtained. Thus, from the perspective of risk management
one requirement would be to store a history of covariance matrices over the period that
the model has been used. Of course, to allow for full back-testing, it will be necessary
to store all the components of the covariance matrix, by which we mean the factor
covariance matrix, the factor exposures and the idiosyncratic covariance matrix. This is
a formidable, but achievable, storage requirement.

Once we have a sequence of portfolio risk forecasts, we will need to assess their efficacy.
Thus, for a given investment house, we will have potentially 100 or so funds, whose risks
have been forecasted over, say, 10 years, at a monthly frequency. At first, this seems a
generous amount of data with which one can test the risk model. However, there are
difficulties, which make this problem much more complicated than it first looks. The first
difficulty, discussed in Chapter 8, is the simple fact that the portfolio weights will keep
changing, even within the month, as some trading will occur, and through time, equity
portfolios will include and exclude stocks as well as changing relatively.

So if we wish to compare the *ex post* volatility with the *ex ante* volatility, as it seems at
first to be natural to do so, we need to store a great deal of information about portfolio
composition through time, and in principle, we should use the beginning of period weights
in the computation of end of period return and risk. In practise, practitioners do not do
this; they consider the fund return and compute the volatility of that.

Another consideration is the unobservable (latent) and changing nature of the volatility process. Financial econometrics seems to be unified in the view that volatility is
a time-varying latent stochastic process. There are numerous models that try to capture its characteristics. We name without references such broad families as generalized
autoregressive, conditional, heteroskedasticity models (GARCH), and stochastic volatility (SV) models. Practitioners have rightly ignored much of this research as very little of it is implementible for a 30,000 stock universe. However, the stochastic latent
nature of volatility means that the matching of *ex ante* and *ex post* becomes especially
problematic.

Suppose we have a volatility (standard deviation) process, which we denote by $\sigma(s)$.
This is a continuous function of time s and is stochastic. In the option pricing literature,
this is frequently modelled by a log-normal Brownian motion with jumps. The cumu-
lative variance from time t to $t+h$ will be $\int_t^{t+h} \sigma^2(s)ds$. If we now consider a return r_t
from time to time $t+1$, we can conceive of it in terms of a constant mean parameter
$\mu r_t = \mu + \int_t^{t+1} \sigma(s)E(s)ds$.

Here, $E(s)$ is some white-noise process with constant unit variance. Suppose, for simplicity that $\mu = 0$, and we compute

$\dot{E}_t(r_t^2)$. This will be the following

$$E_t(r_t^2) = \int_t^{t+1} E_t(\sigma^2(s)) ds.$$

If we now take our *ex post* variance and assume still that $\mu = 0$, then

$$S^2 = \sum_{t=1}^{T} r_t^2 / T$$

and

$$
\begin{aligned}
E_1(s^2) &= \frac{1}{T} \sum_{t=1}^{T} \int_t^{t+1} E_1(\sigma^2(s)) ds \\
&= \frac{1}{T} \int_1^{T+1} E_1(\sigma^2(s)) ds.
\end{aligned}
$$

This tells us that an *ex post* estimator is an unbiased estimator of the expected variance averaged over the period. Whilst this may be reassuring from a statistical perspective, it is unlikely that the *ex ante* forecast from the risk model should equal this number if $\sigma^2(s)$. Write expressions for integral volatility.

6. Diagnostics

In this section, we consider the tools used by the industry for measuring forecast performance. The first one we shall consider is the root mean square error (RMSE). This essentially considers the difference between the *ex ante* and *ex post* values of risk for a particular portfolio. It can be computed both in sample and out of sample. This is a perfectly reasonable number to compute, but the difficulties outlined in the previous section will apply here. Before we detail these difficulties, it may be worth saying a little about RMSE. The theoretical motivation for RMSE stems from the mathematical concept of mean square error, and mean square convergence. We say that $\hat{\theta}_n$ converges in mean square to θ if $\lim_{x \to \infty} E(\hat{\theta}_n - \theta)^2 = 0$.

The key point here is that the concept really applies to the extent to which a sequence of random variables approaches a constant. It requires that we can describe the statistical properties of that sequence, at least in terms of the first two moments. In the context of a risk model, this could be done, but we note that both the changing weights problem and the changing nature of the true but unknown volatility process make this rather challenging.

One procedure that has a great deal to recommend it is what is called the bias test. This is very simple to apply and has the tremendous advantage that it takes the focus off the computation of *ex post* volatility, with all its inherent difficulties as discussed in Section 5. We proceed as follows.

The bias test is a technique used for evaluating the risk model. The explanation is as follows, suppose we want to evaluate the model's ability to forecast risk. We collect a history of returns, $\{r_t, t = 1, \ldots, T\}$, and a history of risk forecasts, $\left\{\hat{\sigma}_{t, t=1,\ldots,T}\right\}$. Note that the return, r_t, is measured over the period t, that is, from time t to time $t+1$, whereas the forecast risk, $\hat{\sigma}_t$, is measured at the start of period t. The standardized return, denoted Z_t, is computed as follows:

$$Z_t = r_t / \sigma_t$$

If the model forecasts risk accurately, then the standard deviation of the time series of $\{Z_t\}$ should be about one. If it is significantly below one, then the model has over-predicted risk. Furthermore, if the standard deviation n of $\{Z_t\}$ is significantly above one, then the model has under-predicted risk. The difficulty in implementing this is the calculation of significance. We discuss this next.

The formal argument is as follows. Under the null hypothesis that the risk model is unbiased, i.e. that the standard deviation of $\{Z_t\}$ is one, and further assuming that the sample size T is large and that the returns are normally distributed, the standard deviation of $\{Z_t\}$ will lie in the interval $[1 - \sqrt{2/T}, 1 + \sqrt{2/T}]$ approximately 95% of the time. This interval represents the 95% confidence interval for the bias statistic. When the computed bias statistic lies within this interval, the null hypothesis that model is unbiased cannot be rejected. Conversely, when the computed bias statistic lies to the right of the interval, we conclude that the model will under-predict risk, and when the bias statistic lies to the left of the interval, we conclude that the model will over predict risk.

Whilst this methodology is very appealing, because it is less focused on the difficulties associated with *ex post* volatility, there are a number of large assumptions. The distribution of the statistic is based upon two critical assumptions. The first is that T is large, so that the estimate of risk will converge in probability to the true risk. This is a standard assumption in the theory of statistics, but we need to think about the realities of the situation. Most risk models are based on effectively 72 monthly observations, either through equal weighting or exponentially weighting a larger number. Thus, the amount of data available may be enough to assume that $\hat{\sigma}_t \rightarrow \sigma_t$. T is the number of scaled returns, often as low as 10. Setting T equal to 72 is probably sufficient if the volatility process is independently and identically distributed (i.i.d.). However, there are large periods of recent history when volatility has been highly auto-correlated, with a consequent loss of information, in which case, T needs to be larger. The second assumption is that returns are normally distributed. This is either true or false, and if the bias test is to be sensibly used, it should be done so jointly with a test for normality. If the returns are non-normal, the critical values (cv) for the bias test will typically move outwards.

A further difficulty with this procedure is that the methodology assumes that the return of the portfolio whose risk is being assessed has a mean of zero. Whilst this may be a reasonable assumption over a very short horizon, it is clearly unreasonable over, say, an annual horizon. If all our calculations are with annualized data, and if the entity in question is likely to have a positive mean, this could bring a substantial bias into proceedings. It should be said though that this bias could be conservative, in the sense that it would make us reject more than accept.

We now suppose that $r_t \sim N(\alpha_t, \alpha_t^2)$ where initially, we shall assume that r_t is i.i.d.
It then follows that $Z_t = \frac{r_t}{\sigma_t} \sim N\left(\frac{\alpha_t}{\sigma_t}, 1\right)$.
And so setting $\theta_t = \frac{\alpha_t}{\sigma_t}$, we see that $Z_t = \theta_t + \varepsilon_t$, where ε_t i.i.d. $N(0, 1)$.

Clearly, $\overline{Z} = \theta_t + \varepsilon_t$, and so $\sum_{t=1}^{T} (Z_t - \overline{Z})^2 = \sum_{t=1}^{T} ((\theta_t - \overline{\theta}) + (\varepsilon_t - \overline{\varepsilon}))^2$. We shall assume that
$\theta_t \sim N(\theta_0, \sigma_\theta^2)$ and independent of ε_t. In which case, $Z_t \sim N(\theta_0, 1 + \sigma_\theta^2)$ and

$$\sum_{t=1}^{T} \frac{\left(Z_t - \overline{Z}\right)^2}{T - 1} \sim (1 + \sigma_\theta^2) \frac{X^2(T-1)}{(T-1)}.$$

Now, without any calculation, we can see that the bias test can be highly inaccurate, as

$$sd = \sqrt{\frac{\Sigma(Z_t - \overline{Z})^2}{T-1}} \sim^2 \sqrt{1 + \sigma_\theta^2} \sqrt{\frac{X^2(T-1)}{T-1}},$$

the upper and lower 2½% cv will be only approximately measured by $1 + 2/\sqrt{T}$ and
$1 - 2/\sqrt{T}$. To get some idea of the magnitudes involved, consider the case where $T = 20$
and $\sigma_\theta = 0$. Here, the exact cv will be 0.68 and 1.31 whilst the approximate cv will be
0.55 and 1.45. For the case, assuming $T = 200, \sigma_\theta = 0$, we get 0.86 and 1.14 for the
approximate values and 0.90 and 1.10 for the exact values. In both cases, we see that
the approximation makes it more likely to not reject bias relative to the true values. If
we add the impact of the variability of the model alpha so that $\sigma_\theta = 5$, this will move the
exact cv upward by 11.2% so that for $T = 20$, the exact values become 0.75 and 1.44.
Whilst this has the effect of virtually equating the upper values of the approximate and
exact procedures, it means that there is a region for the lower value (0.55–0.75) where
the exact test would reject unbiasedness whilst the approximate would not.

7. Time horizons and data frequency

An important issue, not clearly enunciated by many risk model vendors, is the time
horizon of the risk forecast. Implicitly, in the process of making a monthly forecast, and
measuring risk annually, is the idea that the forecast is for 12 months. In litigation, this
can be a point of contention. The reason follows from the well-known fact that if there
is auto-correlation in monthly returns, then annualization by multiplying the monthly
volatility by the square root of twelve can lead to an under- or overestimate, depending
on the sign of the correlation. Thus, the risk manager needs a very clear view of the time
horizon of the forecasts. This is something he or she should require the risk model vendor
to be explicit about. Back-testing cannot be accomplished sensibly without knowing this.

Closely related to the time horizon of the forecast, is the frequency of the model. We
note that for most stocks, prices are available to very high frequencies indeed, likewise,
for yields on gilts and treasuries. However, in the case of fundamental (CS) models, we
do not have the same situation. These models have difficulties in being estimated on
higher frequencies, because much of the data associated with them come out with annual
or biannual accounts. However, for time series models, whose explanatory variables are

themselves prices or returns or yields, there is no such problem. Time series models with macroeconomic factors are quite problematic at a monthly frequency and likely to be very problematic at higher frequencies. This chapter has so far been rather unsympathetic to statistical models. However, they are ideally suited to high-frequency analysis, as they only require prices. The most interesting applications of statistical models that the author has seen has been for intra-day applications.

Now, the great advantage of high-frequency model building and forecast generation is that we can resolve the time horizon issue by building annual measures as sums of weekly measures. These can be bootstrapped or simulated, and some insights into the forecast's tendency to under/over-predict through time aggregation can be gained.

8. The residuals

As with any statistical model, analysis of residuals is an important part of model valida-tion. Using the same notation as before, but allowing the equation below to represent a generic LFM, we can define residuals as follows:

$$\hat{\varepsilon}_{pt} = r_{pt} - \hat{\alpha}_p - \hat{\beta}'_p f_t$$

$$\hat{\alpha}_p = w'\hat{\alpha}; \ \hat{\beta}_p = w'\hat{\beta}; \ \hat{\varepsilon}_{pt} = w'\hat{\varepsilon}_t$$

There are a number of standard and simple exercises that can be carried out with the model residuals. They can be plotted through time to see whether there is any exploitable autocorrelation. Indeed, the autocorrelation functions of the residuals can be computed and standard tests for white noise applied. The factor specification can also be examined.

Many model builders now routinely compute principal components from the residual covariance matrix. The presence of some common factor, not used in the model, will appear in the residual covariance matrix. There also exists more formal procedures for testing the diagonality of a covariance matrix. Indeed, this is a standard problem in classical multivariate analysis. A useful procedure, not used in finance to the author's knowledge, is presented in Muirhead (1982). In particular, he presents, in Theorem 11.2.10, the distribution of a likelihood ratio test for the diagonality of the covariance matrix, assuming normality. It turns out to be asymptotically χ^2, unsurprisingly. However, he or she also presents an extension that allows one to test for diagonality in the presence of elliptical (non-normal) errors.

An issue that arises is the properties of the residuals for the three different types of models. The first two models, time series and statistical, have the attractive property that the sum of the residuals will sum to zero if the models have been estimated by OLS with a constant. Furthermore, under the same assumptions, the sample covariances between factors and residuals will be zero. This considerably simplifies residual analysis.

However, for fundamental (CS) models, the same is not true. It is now the cross-sectional sample mean that is zero. For any given stock, the residuals associated with it will not have a sample time series mean that is zero; it could be any value positive or negative. Whilst this magnitude has the advantage that it can be interpreted as the average stock idiosyncratic alpha, there are disadvantages. It is now necessary to save residuals from all periods, allocating them to stocks, taking care that those stocks were present n the cross-section, and necessarily subtracting the average from them if an estimate of residual variance is required. None of these steps represent major obstacles, but they are all situations in which model users with inadequate statistical training tend to make mistakes.

9. Monte Carlo procedures

The above discussions have demonstrated how truly difficult it is to include real portfolios in an assessment of a risk model. We have discussed many reasons, but the key ones are changing weights and changing volatility. Some of these difficulties can be sidestepped through the use of Monte Carlo methods. Here, there are little practical guidelines as to how to proceed as managers have tended to focus on live or historical portfolios; hence, the following comments are rather exploratory in their nature.

If we look back to the earlier sections of this chapter, we can fully specify stochastic processes that describe the evolution of returns. This can even accommodate time-varying volatility in idiosyncratic returns and factor returns, through the use of vector ARCH and/or SV models. Having specified the returns process with suitable chosen 'true' parameter values, we can imitate the methodology of the risk model vendor by building a risk model from the generated data. This risk model can then be used to build risk forecasts (ex ante risk) for suitably chosen portfolio weights. These forecasts can then be compared with the ex post values of risk, which are actually known in this context, because of the nature of the Monte Carlo experiment.

To be able to do this, we need to know how risk models have been built. This raises an interesting commercial problem as risk model vendors have varying degrees of transparency about their methodologies. So, if a client wants to back-test a vendor model, they need the cooperation of the vendor to help them in carrying out the experiment. Vendors have been traditionally rather secretive about their activities. If the vendor does not cooperate, one can build one's own risk model, either from first principals or through the purchase of risk model building kits, now commercially available. However, the threat of such an action may induce some cooperative behaviour from the vendor, who may be prepared to carry out the exercise himself, making the experiment available to the regulator on demand, the broad results available to the client, but not necessarily revealing key commercial secrets to clients.

10. Conclusions

We have presented a framework for understanding how equity portfolio risk models are built, and how their forecasts are constructed. We have noted a large number of difficulties associated with the validation of such models. Our position is not one of quantitative nihilism. To paraphrase a quotation from a popular film, 'We believe in something'. In particular, we need to set up experiments that compare forecasts with actuals and we need to carry out model analysis that allows us to investigate model misspecification. For the former, Monte Carlo analysis is strongly recommended. For the latter, bias tests and residual analysis seem useful tools. At a practical level, a risk model validation would also involve inspection of the documentation processes, implementation and degree of embeddedness of the model in the risk management, carried out by its user.

References

Brooks, M., Beukes, L., Gardner, D. and Hibbert, J. (June 2002). Predicted Tracking Errors-The Search
    Continues. Investment Risk Working Party, FIA Working Paper. Presented at Faculty and Institute of
    Actuaries Finance and Investment Conference.
Campbell, J., Lo, A. and MacKinlay, C. (1997). *The Econometrics of Financial Markets*. Princeton University
    Press, New Jersey.
Connor, G. (1995). The 3 types of factor models: A comparison of their explanatory power. *Financial Analysts
    Journal*, 51(3), 42–6.
Dhrymes, P.J. (1970). *Econometrics*. New-York: Harper and Row.
Gardner, D., Bowie, D., Brooks, M., *et al.* (2000). Predicted Tracking Errors, Fact or Fantasy? *Portfolio
    Risk and Performance Working Party, Financial and Institute of Actuaries*, presented to FIA, Investment
    Conference.
Muirhead, R.J. (1982). *Aspects of Multivariate Statistical Theory*. J. Wiley & Sons Inc.

10 Dynamic risk analysis and risk model evaluation

Günter Schwarz* and Christoph Kessler*

Abstract

In this article, we present explorative methods for measuring investment risk. While the underlying concepts are purely quantitative, the way results are visualized and interpreted is rather qualitative but nevertheless rigorous. The methods are in particular well suited for the task of evaluating the performance of investment risk models that are used in the investment decision process of an asset manager, where the time horizon is months and not days. Such a model cannot be subjected to rigorous statistical tests as the amount of time required to achieve significance would be far too long. The explorative methods described in this chapter, however, provide 'immediate' results as they visualize the dynamics of risk behaviour 'in real time', allowing direct comparison with what the risk model anticipated at the same time. The chapter begins with a brief discussion of in-sample tests versus out-of-sample tests used in investment risk measurement and how these might be utilized in model validation. We then introduce the underlying concepts of cumulative variance and covariance. We outline an approach of viewing risk over time that does not enforce a pre-determined fixed time horizon and apply it in various examples. These examples cover general risk measures like beta, total and active risk that can be applied for either single securities or portfolios. Additional risk measures that are standardized by the risk forecast are then applied to evaluate risk model out-of-sample performance. Finally, we discuss some criteria for judging the quality of the risk model.

1. Introduction

Investment risk models are widely utilized in the finance industry. The scope and shape of these models are as diverse as the way they contribute to investment or trading decisions. One extreme might be an option trader who requires a model that describes the anticipated distribution of the returns of a small number of parameters as accurate as possible to be competitive in the market. In turn, a policy risk tool to be used in asset/liability management for pension funds much more focuses on the long-term covariance between aggregated markets and asset classes.

The ways the performance of such models is assessed are as diverse as the models' scopes and structures and in particular the intended time horizon. Although it might be simple to check the performance of an option trader's model by measuring the accuracy with which the trader is able to react in the market, it is about impossible to confirm

* UBS Global Asset Management, Global Investment Solutions, Zurich, Switzerland.

the quality of a model with a horizon of decades as one would need to wait hundreds of years until one has statistically significant results.

In general, models can be checked in-sample or out-of-sample. In-sample checks are based on the same data that were originally used to estimate the model. Their main purpose is to evaluate whether the model is rich enough to capture the targeted risk characteristics without being 'over-fitted'. In the context of linear factor models that are commonly used in the finance industry, important questions are as follows:

- Does the model have enough factors to capture the correlations of the assets, that is, can assets' specific risks be regarded as being uncorrelated without distorting the risk geometry?
- Are the factors adequate, that is, do they facilitate the management and allocation of risk in the investment process?
- Are the assets' factor exposures well chosen and/or estimated?
- Is the factor explained risk reasonable in size and pattern across the whole universe of securities?
- Can the model be made more parsimonious without loss of precision?

Let us just mention that these questions can be answered by testing the model on a set of real or artificial (simulated) portfolios. Thoroughly designed in-sample tests give a good sense of the behaviour of the model and are the basis for the decision whether the model is suited for being made operational.

In contrast to this, out-of-sample checks evaluate the model on data not already used for model calibration. In particular, the interest is on the forward-looking side, that is, evaluation with data from a period that goes beyond the estimation time frame. Whenever real-life risk management applications are evaluated, two effects need to be taken into account: the underlying security and market risk is changing over time and the portfolio structure may change over time as well.

When dealing with out-of-sample tests, we distinguish between the following general methods:

- The *static approach* evaluates the model over a pre-determined out-of-sample period, for example, on a yearly basis since introduction of the model.
- The *semi-static approach* uses a rolling window of pre-determined length, for example, a 1-month window over a year.
- The *dynamic approach* uses explorative statistics and charts that allow evaluating the dynamic ex post risk behaviour of the assets and portfolios over an extended longer period (e.g., since introduction of the model) independent of a fixed window size or frequency. This then can be compared with the numbers anticipated by the model over the same period. In addition, the dynamic approach allows analyzing how the model worked over any sub-period of the out-of-sample period. Model changes can be accommodated as easily as interim changes in portfolio structure.

In this chapter, we give a detailed description of the dynamic approach for out-of-sample analysis of risk behaviour.

2. Volatility over time and the cumulative variance

2.1. Theoretical background

There exist different ways of looking at the variation of volatility over time. For example, it is possible to estimate volatility using a fixed number of observations over a pre-determined time span and move this estimation window forward over time, or one can use generalized autoregressive conditional heteroskedasticity (GARCH) modelling. For an introduction see for example Jorion or McNeil, Frey Embrechts. Here, we suggest an approach that goes back to the original definition of volatility as the square root of the sum of squared returns and utilizes the underlying process of the cumulated sums of squared returns – the so-called quadratic variation process. Notice that the actual instantaneous variance can be seen as the first derivative of this process.

Given returns $R(s)$ for $s \geq 0$, let the *cumulative realized variance* process be defined as

$$\text{CumVar}(t) := \sum_{0 < s \leq t} R^2(s), \quad t \geq 0. \tag{10.1}$$

The *realized variance* over the time period from t up to T is then given by

$$\text{CumVar}(t, T) := \sum_{t < s \leq T} R^2(s) = \text{CumVar}(T) - \text{CumVar}(t). \tag{10.2}$$

So the accumulated variance over a period equals the increment of the cumulative variance. With time measured in years, the slope of the cumulative variance over the time period t up to T equals the conventional variance estimate under the assumption of zero means. So, if dt denotes the observation frequency [resulting in a number of $(T - t)/dt$ return observations in the time window t to T], and using $\hat{\sigma}^2_{dt}$ and $\hat{\sigma}^2$ for the dt frequency variance estimator and the annualized variance estimator, respectively, we have

$$\frac{\text{CumVar}(t, T)}{T - t} = (T - t)^{-1} \cdot \sum_{t < s \leq T} R^2(s) = \frac{\hat{\sigma}^2_{dt}(t, T)}{dt} = \hat{\sigma}^2(t, T). \tag{10.3}$$

If returns have a zero mean, then the realized variance is an unbiased estimate of the cumulated variance over the period from t up to T: given the variance of the returns $R(s)$,

$$\sigma^2(s) := E\left(R^2(s)\right), \tag{10.4}$$

and the cumulative variance of the returns over the period t up to T is

$$\sigma^2(t, T) := \sum_{t < s \leq T} \sigma^2(s), \tag{10.5}$$

then the expected realized variance coincides with the variance, that is,

$$E(\text{CumVar}(t, T)) = \sigma^2(t, T). \tag{10.6}$$

Sometimes the cumulative variance is also called the integrated (squared) volatility. See for example Barndorf-Nielsen, Shephard. Although our analysis is, strictly speaking, about the second moments and not about the variance, for most practical applications in finance with frequencies up to monthly data, the mean is close enough to zero to justify neglecting it. In fact, volatility estimates derived this way are often more stable.

If in addition the variance or volatility is constant over the time period from t up to T, then the realized variance is an unbiased estimate of this underlying volatility, that is,

$$\sigma^2(t, T) = \sum_{t < s \leq T} \sigma^2(s) = \sigma^2(T - t) \tag{10.7}$$

and

$$E(\mathrm{CumVar}(t, T)) = \sigma^2(T - t). \tag{10.8}$$

So with time measured in years, the expected slope of the cumulative variance over the time period t up to T equals the square of the annualized volatility σ, that is,

$$E\left(\frac{\mathrm{CumVar}(t, T)}{T - t}\right) = \sigma^2. \tag{10.9}$$

In particular, if we assume that volatility is constant over time, then the cumulative variance fluctuates around the *constant variance line* given by

$$v(t) = \sigma^2 \cdot t, \quad \text{for } t \geq 0. \tag{10.10}$$

How large is the deviation of the realized cumulative variance relative to the underlying cumulative variance curve? In case of uncorrelated squared returns over time (i.e., zero autocorrelation), the standard deviation of the realized variance can be calculated easily:

$$\mathrm{Var}(\mathrm{CumVar}(t, T)) = E\left[\mathrm{CumVar}(t, T) - \sigma^2(t, T)\right]^2 = \sum_{t < s \leq T} \left[E\left(R^4(s)\right) - \sigma^4(s)\right]. \tag{10.11}$$

In case of normally distributed returns, this reduces further to

$$\mathrm{Var}(\mathrm{CumVar}(t, T)) = 2 \sum_{t < s \leq T} \sigma^4(s). \tag{10.12}$$

Finally, if volatility is constant, that is,

$$\sigma^2(s) = \sigma^2 \cdot dt, \tag{10.13}$$

then the standard deviation of the cumulative variance equals

$$\mathrm{StDev}(\mathrm{CumVar}(t, T)) = \sigma^2 \cdot \sqrt{2 \cdot dt} \cdot \sqrt{T - t}. \tag{10.14}$$

In this case, the estimation error is proportional to the square root of time. The conventional annualized variance estimate under zero mean assumption equals the slope of the realized cumulative variance, and the estimation error is given by

$$\text{StDev}\left(\frac{\text{CumVar}(t, T)}{T-t}\right) = \sigma^2 \left(\frac{\sqrt{2 \cdot dt}}{\sqrt{T-t}}\right). \tag{10.15}$$

If one looks at the slope measured in volatility units instead of variance units, then a Taylor expansion of the square root function yields

$$\text{StDev}\left(\frac{\sqrt{\text{CumVar}(t, T)}}{\sqrt{T-t}}\right) \approx \sigma \left(\frac{\sqrt{dt}}{\sqrt{2(T-t)}}\right) = \frac{\sigma}{\sqrt{2 \cdot N(t, T)}}. \tag{10.16}$$

So the relative error depends purely on the number of observations, $N(t, T) = (T-t)/dt$ in the time period from t to T. For later reference, Table 10.1 shows the estimation error for different popular data frequencies and time spans. For the sake of completeness, let us further mention the well-known fact that under the assumption of independent, identically and normally distributed returns, the variance estimate is χ^2 distributed. In particular, the estimation error distribution is skewed, and the standard deviation has to be used with this in mind.

Table 10.1 Volatility estimation error

Observation Frequency	Observation Period Length				
	1 Month	1 Quarter	1 Year	2 Years	5 Years
Daily	15	0.09	0.04	0.03	0.02
Weekly	34	0.20	0.10	0.07	0.04
Monthly	71	0.41	0.20	0.14	0.09

The standard deviation of $\hat{\sigma}/\sigma$ equals $\sqrt{dt}/\sqrt{2(T-t)}$ or $1/\sqrt{2 \cdot N(t, T)}$ based on return frequency dt, observation period length $(T-t)$ and number of observations $N(t, T)$

2.2. Explorative approach

As the concept of cumulative variance is rather simple, its explorative use can be quite fruitful. In this section, artificial data will be used to explain how the concept is applied. The simplest example to start with consists of randomly generated returns with constant volatility. Daily returns are drawn independently from a normal distribution with zero mean and a constant volatility of 10% on an annualized basis. Assuming 260 trading days in a year, daily volatility is 10% divided by the square root of 260 or 0.6%.

Figure 10.1 plots the realized cumulative variance for three random return series. The figure is complemented with constant volatility lines that make it possible to 'guesstimate'

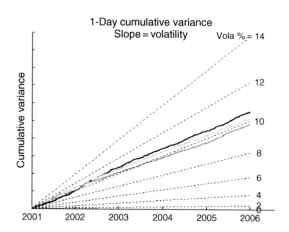

Figure 10.1 Realized cumulative variance based on normally distributed daily returns with 10% volatility annualized

the volatility just by visual inspection – the cumulative variance should in the example move in parallel to the 10% volatility line. As can be seen, this works fine if daily data is used. The variation in slope over sub-periods depends on the length of the sub-period. For the standard deviation errors of these slopes, see Table 10.1. Of course, if one looks at time periods with particularly high volatility, the estimation error is likely to be higher than the figures suggested in Table 10.1 as in this case, one is looking at what is called a maximally selected statistics. This is a well-known effect that is dealt with in sequential statistical analysis.

For practical purposes, it is important to get a sense for how precise the realized cumulative variance reflects the underlying variance. While the estimation errors in Table 10.1 give a perspective on this issue and Figure 10.1 indicates results for daily data, the next item would be to have a look at the impact of data frequency. In Figure 10.2, the realized

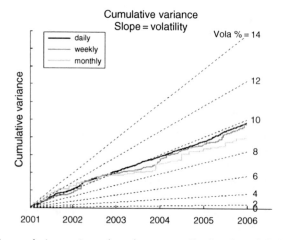

Figure 10.2 Realized cumulative variance based on normally distributed daily, weekly and monthly returns with 10% volatility annualized

cumulative variance is shown for a single underlying data series but now using daily, weekly and monthly returns for calculating the cumulative variance.

As an example, the monthly volatility estimate over the full 5-year period is close to 9% and under-estimates the true volatility by 10% in relative terms. This is just in line with the 10% estimation error from Table 10.1. On a yearly basis, one could see periods with volatility as low as 5% or going up to 12%. Table 10.1 would indicate a relative estimation error of 20%, that is, estimates between 8% and 12%. This 'discrepancy' is due to the effect of looking at the extremes over a given time period. Of course, it would be possible to exploit this further by sampling the distribution of the maximally selected slopes over certain periods of time.

As the estimation error only depends on the number of observations, the statistical behaviour of the slope of the cumulative variance (i.e. the estimated variance) is roughly the same within each of the following three examples:

1. around 12 observations: 2-week periods based on daily data, quarterly periods based on weekly data or yearly periods based on monthly data,
2. around 25 observations: monthly periods based on daily data, half-year periods based on weekly data or 2-year periods based on monthly data and
3. around 50–60 observations: quarterly periods based on daily data, yearly periods based on weekly data or 5-year periods based on monthly data.

The next example resembles a regime shift in the underlying volatility. We assume that during the first half of a 5-year period, annualized volatility is at 10% and then suddenly rises to 15%. Figure 10.3 shows the daily, weekly, and monthly cumulative variance based on the same underlying daily return series. While with monthly data it is really hard to guess the underlying volatility from the chart, the weekly cumulative variance already indicates the regime shift pattern, and the daily data series quite nicely picks up the actual regime shift.

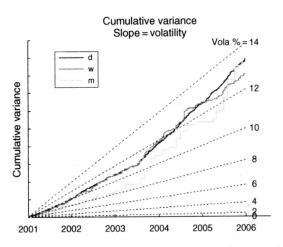

Figure 10.3 Realized cumulative variance based on normally distributed daily, weekly and monthly returns with 10% volatility during the first 2.5 years and 15% during the second 2.5 years

In the third example, we assume a volatility shock, which is a quite common phenomenon in financial markets. This is another interesting scenario indicating how the cumulative variance behaves (Figure 10.4).

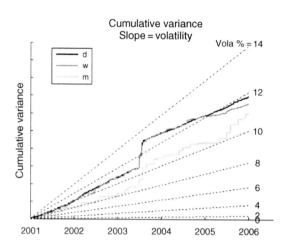

Figure 10.4 Realized cumulative variance based on normally distributed daily, weekly and monthly returns with 10% constant volatility, apart from a 1-month spike in volatility of 40% in the middle of the 5-year period

It is hard to extract the exact magnitude and length of the volatility spike from the daily and weekly cumulative variance chart. However, one can at least nicely see the pattern: stable volatility – spike in volatility – going back to the original volatility level before the spike. Monthly data, however, are not even sufficient for this conclusion.

Finally, it is interesting to compare the explorative characteristics of the cumulative variance with commonly used methods like rolling window volatility estimates. The three volatility scenarios from above – 'constant volatility', 'volatility regime shift' and 'volatility shock' – will be used to indicate the differences in behaviour.

In Figure 10.5, the daily and the 5-day realized cumulative variance is shown. The 5-day cumulative variance is based on overlapping 5 trading-day intervals, that is, Monday to Monday, Tuesday to Tuesday.... Overlapping 5-day returns are quite useful as a compromise when analyzing global assets and portfolios that span different time zones.

The realized cumulative variance based on 5-day overlapped data also does as well indicate the underlying true cumulative variance. Figure 10.6 shows the corresponding charts for rolling window estimates with window sizes 1 month, 1 quarter and 1 year based on daily return observations. The left-hand side covers the 'regime shift' scenario and the right-hand side the 'volatility shock' scenario.

For the regime shift shown on the left-hand side of Figure 10.6, the 1-month window estimate picks up the increase rather quickly, but is in fact too noisy. The 3-month estimate does a good job, but it takes 3 months to end up with the correct estimate. For the smooth yearly estimate, it takes a year of 'rising volatility' to adjust to the new regime. The comparison with Figure 10.5 nicely shows the advantage of the concept of cumulative variance.

For the volatility shock, the peak of the 1-month window estimate picks up the underlying volatility shock due to the fact that the shock did last for just a month. But it is

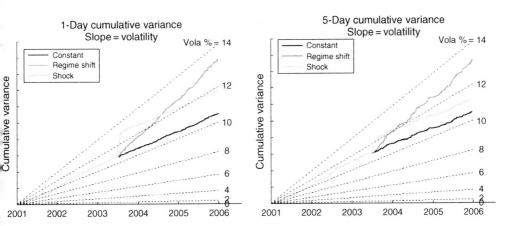

Figure 10.5 Realized cumulative variance based on normally distributed daily (left-hand side) and 5-day overlapping (right-hand side) returns with 'constant volatility' (at 10%), 'volatility regime shift' (from 10% to 15% after 2.5 years) and 'volatility shock' (10% volatility with 40% spike after 2.5 years for a month)

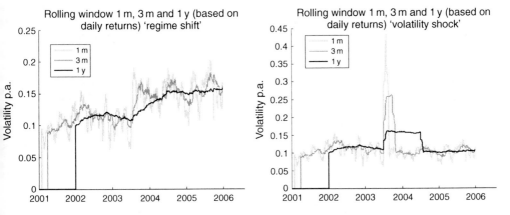

Figure 10.6 Realized volatility over fixed size rolling windows with length 1 month, 1 quarter and 1 year. Estimates are based on daily returns. The underlying scenarios are 'volatility regime shift' (from 10% to 15% after 2.5 years) on the left-hand side and 'volatility shock' (10% volatility with 40% spike after 2.5 years for a month) on the right-hand side

hard to guess from this chart that it was a 1-month shock at a constant volatility of 40%. This is different for the cumulative variance in Figure 10.5 (left-hand side). The pattern is more or less obvious there, and a separate evaluation of the high slope (high volatility period) would even yield the 40-% estimate.

Of course these examples have been artificial, and are therefore no proof of the usefulness of the cumulative variance approach in general. However they already indicate the advantages compared to commonly used methods.

2.3. *Practical application for portfolio total and active risk*

The realized cumulative variance can be used to explore the dynamics of risk for single assets, but as well for benchmarks, portfolios and portfolios relative to benchmarks. As discussed in the previous paragraphs, the time variation in risk can be assessed over a given time period, provided there is a sufficient amount of data. So, the desired time resolution determines the required data frequency.

The first example is a European equity fund with the Morgan Stanley Capital International (MSCI) Europe index as benchmark. We draw the realized cumulative variances for the portfolio, the benchmark and the portfolio versus the benchmark based on daily returns. Figure 10.7 gives a good overview about the time variation in risk. For portfolio and benchmark, there is a high volatility of 30% per annum in the period preceding the Iraq war in the first quarter of 2003. Then both portfolio and benchmark risks go down immediately to 14% and continue to drop to 10–12% until the May 2006 shock, which was passed with a volatility of ca. 25%. After May 2006, risks are back at low levels of below 10%. The active or relative risk of the portfolio versus the benchmark is around 5% in the first quarter of 2003. Afterwards, there is not much fluctuation, and the active risk is quite stable at 2.5% with only a minor impact of the May 2006 shock. After May 2006, the active risk is low at 2%.

Of course, one could have also come to this conclusion by looking at a set of rolling fixed-length volatility estimates with varying length and GARCH estimates. The difference, however, is that with a cumulative variance plot, this can be seen in a single chart that in addition is very simple to calculate from portfolio and benchmark performance data.

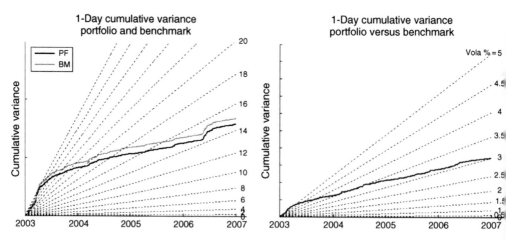

Figure 10.7 Realized cumulative variance for a European equity portfolio with MSCI Europe benchmark based on daily returns

The next example is a global equity portfolio with the MSCI World as benchmark. In this case, it is adequate to use 5-day overlapped returns to counter the effect of the different time zones. Daily returns would lead to a downward bias in risk figures because of the low autocorrelation within the day. Even allowing for the slightly reduced precision compared to the above example due to the 5-day overlapped returns, the cumulative variance still does a reasonable job in showing the dynamic behaviour of risk. Please notice that the time series start at the end of May 2003 and do not include the pre-Iraq war period (Figure 10.8).

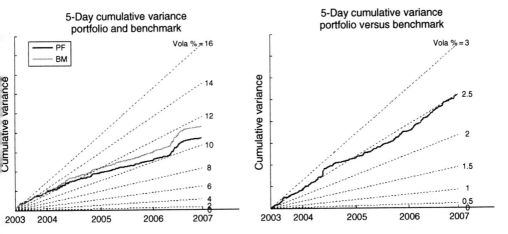

Figure 10.8 Realized cumulative variance for a global equity portfolio with MSCI World benchmark based on 5-day overlapped returns

3. Beta over time and cumulative covariance

3.1. Theoretical Background

The concept of the cumulative variance can be adapted to the covariance in a straightforward fashion, for example, the covariance of a portfolio versus its benchmark or assets versus the underlying market indexes. However, here we will focus on beta, that is, the market exposure of a portfolio or asset in general instead, as this introduces an additional explorative concept.

The beta of a portfolio (or asset) with returns $R_P(t)$ for $t \geq 0$ versus a benchmark (or market) with returns $R_B(t)$ for $t \geq 0$ is defined as the covariance of portfolio returns versus benchmark returns divided by the variance of the benchmark returns, that is,

$$\beta_{P,B}(t) := \frac{\text{Cov}(R_P(t), R_B(t))}{\text{Var}(R_B(t))}. \tag{10.17}$$

Let us generalize the concept of the cumulative variance to that of the *cumulative realized covariance* process, which can be defined as

$$\text{CumCovar}_{P,B}(t) := \sum_{0 < s \leq t} R_P(s) \cdot R_B(s), \quad t \geq 0. \tag{10.18}$$

The *realized covariance* over the time period from t up to T is then given by

$$\text{CumCovar}_{P,B}(t, T) := \sum_{t < s \leq T} R_P(s) \cdot R_P(s) = \text{CumCovar}_{P,B}(T) - \text{CumCovar}_{P,B}(t) \tag{10.19}$$

The accumulated covariance over a period equals the increment of the cumulative covariance. Furthermore, the *realized beta* over the time period t up to T can be defined as

$$\hat{\beta}_{P,B}(t, T) := \frac{\text{CumCovar}_{P,B}(t, T)}{\text{CumVar}(t, T)}. \tag{10.20}$$

Let us mention that if the returns have zero means, then this is an unbiased estimate, that is,

$$E\left(\hat{\beta}_{P,B}(t, T) \,|\, R_B(t < s \le T)\right) = \sum_{t < s \le T} \beta_{P,B}(s) \cdot \frac{R_B^2(s)}{\sum_{t < u \le T} R_B^2(u)}. \tag{10.21}$$

This means that the conditional expectation of the realized beta given the benchmark returns is a weighted average of the betas. The corresponding weights are determined by the squares of the benchmark returns. We will come back to this expression later. If beta is constant over the time period, then the right-hand side of Equation 10.21 reduces to the respective constant beta.

The conditional variance of the realized beta given the benchmark returns can be derived as

$$\text{Var}\left(\hat{\beta}_{P,B}(t, T) \,|\, R_B(t < s \le T)\right) = \frac{1}{\sum_{t < u \le T} R_B^2(u)} \cdot \sum_{t < s \le T} \sigma_{P,B}^2(s) \cdot \frac{R_B^2(s)}{\sum_{t < u \le T} R_B^2(u)}, \tag{10.22}$$

with the *residual variance* defined as

$$\sigma_{P,B}^2(s) := \text{Var}\left(R_P(s) - \beta_{P,B}(s) \cdot R_B(s)\right). \tag{10.23}$$

This is an expression similar to that for the conditional mean: the conditional variance is a weighted average of the residual variances, where the weights are again determined by the squares of the benchmark returns. In case of constant residual variance, the variance of the realized beta can be simplified to the well-known formula

$$\text{Var}\left(\hat{\beta}_{P,B}(t, T) \,|\, R_B(t < s \le T)\right) = \frac{1}{N(t, T)} \cdot \frac{\sigma_{P,B}^2}{\hat{\sigma}_B^2(t, T)}, \tag{10.24}$$

where $N(t, T)$ is the number of observations in the period of t up to T, and

$$\hat{\sigma}_B^2(t, T) := \frac{1}{N(t, T)} \cdot \sum_{0 < s \le T} R_B^2(s) \tag{10.25}$$

is the realized benchmark variance. The estimation error therefore depends on the ratio of residual risk to benchmark risk and is proportional to the reciprocal of the square root of the number of observations.

For practical applications, the ratio of residual volatility to benchmark volatility is in general lower than 1 for 'long only' portfolios and often between 25% and 50%. For

Table 10.2 Beta estimation error

Observation frequency	Observation period length				
	1 Month	1 Quarter	1 Year	2 Years	5 Years
Daily	0.21	0.12	0.06	0.04	0.03
Weekly	0.48	0.28	0.14	0.10	0.06
Monthly	1.00	0.58	0.29	0.20	0.13

The standard deviation of $\hat{\beta}_{P,B}/(\sigma_{P,B}/\sigma_B)$ equals $\sqrt{dt}/\sqrt{T-t}$ or $1/\sqrt{N(t,T)}$ based on return frequency dt, observation period length $(T-t)$ and number of observations $N(t,T)$ (Table 10.2).

assets, the respective ratio can vary substantially, and for equities, it can sometimes be as low as 50%, but in most cases, it falls between 100% and 300%. The numbers shown in Table 10.2 need to be multiplied by this ratio to get to the estimation error of beta.

3.2. Explorative approach

By definition, realized beta is the ratio of the cumulative covariance of the portfolio (or asset) versus the benchmark (or the market) and the benchmark (or market) cumulative variance. This leads to the idea of creating a chart where one plots the cumulative benchmark variance on the horizontal axis and the cumulative covariance on the vertical axis. Then, the slope of the resulting curve over a particular time period equals the realized beta over this time period. This way, it is possible to visualize the dynamics of beta over time within a single chart, that is, indicate the beta variation in time over any sub-period or the full period. So we can utilize the same explorative framework as for the cumulative variance.

Figure 10.9 is an example using randomly generated returns. Both portfolio and benchmark returns are normally distributed, portfolio beta equals 1 and the ratio of residual to

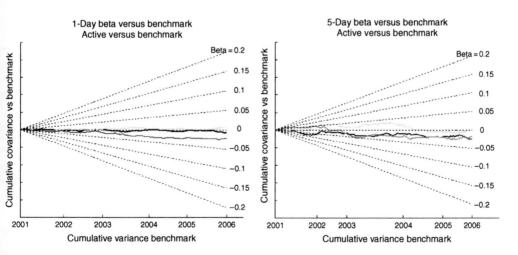

Figure 10.9 Realized cumulative active portfolio beta based on normally distributed daily returns (left-hand side) and 5-day returns (right-hand side). Active beta is zero and the ratio of residual to benchmark volatility is 1/3

benchmark volatility is 1/3. We focus on the active beta (which equals beta minus 1) and the active portfolio returns (defined as the difference between portfolio and benchmark returns) rather than the corresponding portfolio figures to get a good resolution of the cumulative beta chart. So the vertical axis shows the covariance of the active portfolio returns versus the benchmark returns. On the horizontal axis, time is 'running' according to the cumulative benchmark variance. On the left-hand side of Figure 10.9, the effect of this time transformation is hardly visible as benchmark volatility is constant, and we use daily data. However, for the weekly data sample on the right-hand side of Figure 10.9, the effect of this time scale can be seen clearly as the random fluctuation of the benchmark cumulative variance becomes larger. Over any chosen time period, the slope of the covariance curve equals the realized active beta over that period. Let us mention the relation to Equation 10.21, which shows how the time running with the benchmark cumulative variance determines the weighting for the conditional expectation of beta given the benchmark returns.

The estimation error can be derived from the 'daily' or 'weekly' column of Table 10.2 together with the ratio of residual to benchmark risk, which is 1/3 in this example (i.e., by dividing the values in the table by 3). For example, for the 5-year period, the daily beta error is 1%, and for 1-year periods, it is 2%. The weekly estimation error is 2% for the 5-year period or 5% for a 1-year period. With weekly data, the cumulative beta over 1-year periods is already quite noisy. However, there are applications in the financial industry with only monthly data availability. Figure 10.10 illustrates the lack of precision one has to accept in such a case.

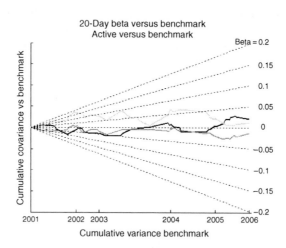

Figure 10.10 Realized cumulative active beta based on normally distributed monthly returns. Active beta is zero and the ratio of residual to benchmark volatility is 1/3

The next example indicates the effects of shifts in the underlying beta. For the first 2.5 years, beta equals 1, for one quarter it peaks at 1.5 and for the rest of time beta is at 0.95. Figure 10.11 shows the plot for three samples of daily and 5-day overlapping data. In both plots, the underlying changes in beta can be extracted from the graph.

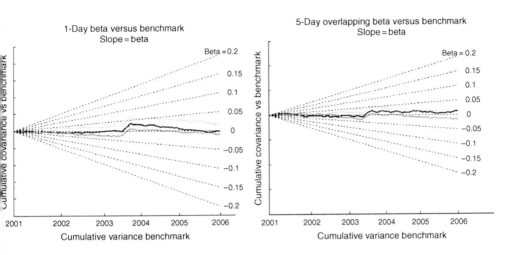

Figure 10.11 Realized cumulative active beta based on normally distributed daily returns (left-hand side) and 5-day overlapping returns (right-hand side) with active beta 0% for the first 2.5 years, then for a quarter a sudden increase to 50% and for the rest of time −5%

3.3. Practical application for portfolio beta

The cumulative beta charts can be a useful tool for checking the dynamics of portfolio beta. Figure 10.12 plots the active (portfolio minus benchmark) cumulative beta for a European portfolio versus the MSCI Europe as benchmark and for a global portfolio versus the MSCI World as benchmark. We use daily returns for the European portfolio and 5-day overlapping returns for the global portfolio.

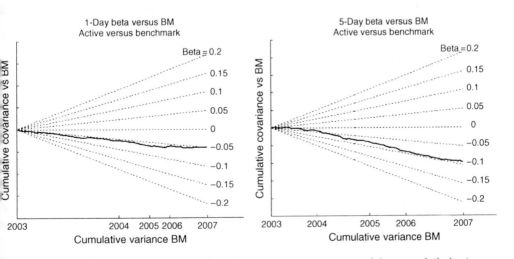

Figure 10.12 Realized cumulative active beta for a European equity portfolio on a daily basis on the left-hand side and for a global equity portfolio with 5-day overlapping returns on the right-hand side

In particular, for the Europe portfolio, one can see the impact of the time scaling introduced by the cumulative benchmark variance. In 2003, there was a high benchmark volatility causing the time to expand accordingly. For both portfolios and benchmarks, the behaviour of risk is shown in Figures 10.7 and 10.8. For the European portfolio, the active beta ranged between -5% and -10% before 2006, and then in 2006, it was close to zero. The global portfolio had an active beta of approximately -5% in 2004, this then decreased to about -15% and later increased slightly to -10%. At the end of the observation period, it was back at -5%, with a spike of -15% right at the end of the period.

As one could see not only from these two real-life portfolios but also from the examples based on simulated data, the cumulative beta charts provide a useful tool for analyzing beta over time. Of course, precision depends on both the data frequency and the ratio of residual risk to benchmark risk.

4. Dynamic risk model evaluation

4.1. Theoretical background

Based on the simple theory derived in the above sections, we suggest two approaches for an explorative risk model evaluation:

1. Add the respective ex ante forecasts to the cumulative variance and beta charts.
2. Explore the cumulative variance of the returns standardized by the risk forecast.

For the first approach, we use a single plot to compare the realized cumulative variance

$$\text{CumVar}_R(t) := \sum_{0 < s \leq t} R^2(s), \quad t \geq 0. \tag{10.26}$$

with the *forecasted cumulated variance*

$$\text{CumFVar}_R(t) := \sum_{0 < s \leq t} \sigma^2_{F;R}(s), \quad t \geq 0. \tag{10.27}$$

Here, $\sigma^2_{F;R}(t)$ denotes the risk forecast at time t for the next day. The same can be done for the beta or factor exposure. Here, we plot in a single chart on the vertical axis both the realized cumulative covariance

$$\text{CumCovar}_{P,B}(t) := \sum_{0 < s \leq t} R_P(s) \cdot R_B(s), \quad t \geq 0 \tag{10.28}$$

and the *forecasted realized covariance* based on the beta forecast $\beta_{F;P,B}(t)$, that is,

$$\text{CumFCovar}_{P,B}(t) := \sum_{0 < s \leq t} \beta_{F;P,B}(s) \cdot R_B(s), \quad t \geq 0. \tag{10.29}$$

In this plot, the time on the horizontal axis is running with the realized cumulative benchmark variance.

The second approach is to look at the cumulative-standardized variance, that is, base the analysis on the realized return divided by the risk forecast

$$\text{CumVar}_{\bar{R}}(t) := \sum_{0 < s \leq t} \left(\frac{R(s)}{\sigma_{F;R}(s)} \right)^2, \quad t \geq 0. \tag{10.30}$$

If the forecasts are perfect, then the standardized returns have unit volatility. If in particular volatility is constant, the cumulative variance follows a χ^2 distribution, and the statistics for constant volatility as outlined in Section 2.1 can be applied to check the assumption that the risk forecasts are unbiased.

4.2. Realized versus forecast cumulative variance

We proceed with the Europe portfolio example. Figure 10.13 shows cumulative total, benchmark and active variance realized as in Figure 10.7 plus the cumulated variances based on risk model forecast as shown in Figure 10.14. The risk model under consideration is a long-term risk model and as such does not reflect the short-term realized volatility pattern. In times of high volatility, the model underestimated the total and benchmark risk, whereas in low volatility times, it substantially overestimated these values. For the relative risk of the portfolio relative to the benchmark, the situation looks slightly different. The model performed better but still had a bias to overestimate risk. It is interesting to see how the cumulative plots 'smoothes out' the risk over time.

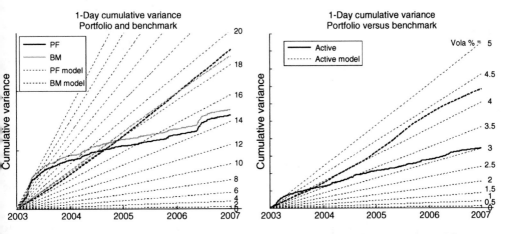

Figure 10.13 Europe portfolio total, benchmark and active risk realized versus risk model

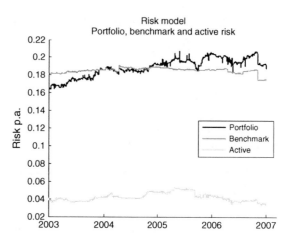

Figure 10.14 Europe portfolio total, benchmark and active risk based on risk model

4.3. Realized versus forecast cumulative beta

Looking at the same European portfolio using the same long-term risk model, Figure 10.1? shows the cumulative realized and forecasted active beta on the left-hand side and the forecasted risk model active beta on the right-hand side. The model in this case does a reasonable job, only slightly overestimating the magnitude of beta. For example, the model beta was showing 5% in 2006, whereas the realized beta was closer to zero. The model beta also behaves more smoothly than the realized beta.

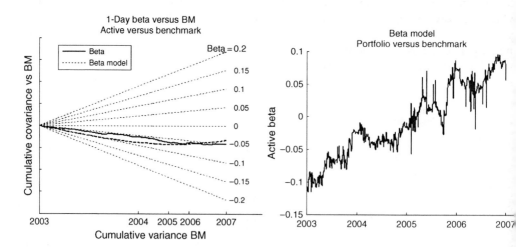

Figure 10.15 Europe portfolio active beta realized versus risk model

4.4. Cumulative variance of forecast-standardized returns

Our second approach to judge the performance or possible bias of the risk model is to look at the cumulative variance of the standardized returns, where the standardization is given by the risk model forecast. In Figure 10.16 on the left-hand side, once again the European portfolio is shown with daily data using the long-term risk model for standardizing returns. On the right-hand side for a shorter period of time, the global portfolio already discussed above is shown with a shorter-term risk model used for standardization. The slope of the cumulative variance chart indicates the performance of the risk model. Whenever the slope over a time period is above 1, then the model under-estimates risk over this time period. A slope below 1 indicates that the model over-estimates risk.

On the left-hand side for the European portfolio, the pattern for portfolio and benchmark risk is identical while the active risk differs mostly in the first quarter of 2003. Apart from May 2006, risk has been ca. 40% over-estimated. The active risk figures are slightly less biased than the portfolio and benchmark figures as the active slope is slightly closer to one.

On the right-hand side for the global portfolio, 5-day overlapping returns are used. Again, the active risk forecast is less biased than the portfolio and benchmark risk. Portfolio and benchmark forecast quality is quite similar. During May 2006, the slope was at 1.4, and the model under-estimated risk accordingly in this period. Before May 2006, the slope was between 0.6 and 1.0, which means the model first over-estimated risk substantially, and then this bias was going down to zero. In the second half of 2006, the model again over-estimated risk. Over the full 2-year period, the model was unbiased. The model performed reasonably well for active risk, over-estimating by ca. 30% in the last quarter of 2005 and the first quarter of 2006. During May 2006, the model forecast was unbiased (Figure 10.16).

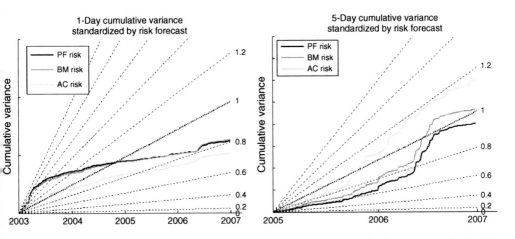

Figure 10.16 Europe (left-hand side) and global portfolio (right-hand side) forecast-standardized cumulative variance for portfolio, benchmark and portfolio versus benchmark returns

5. Summary

Cumulative variance and beta plots offer a great potential in analyzing the time variation of risk numbers in a single plot. By just visualizing the evolution of the cumulative variance or beta, it is possible to replace a whole series of rolling window or GARCH estimates that would otherwise be needed to carry the same information. In addition, the suggested plots are very easy to generate and intuitive to interpret.

In this chapter, we presented ideas of how to apply these plots in the context of a dynamic evaluation of investment risk models. By standardizing realized risk figures by model-estimated figures, model performance evaluation can be done within the scope of a single chart. Standardization takes care of both the effect of risk changing over time and of any portfolio restructuring done during the analysis, whereas the technique of the cumulative plot makes this result available 'instantaneously'.

The method can easily be extended at other concepts of risk, like correlation analysis or multivariate exposures.

References

Barndorf-Nielsen, O.E. and Shephard, N. (2002) Econometric analysis of realized volatility and its use in estimating stochastic volatility models. *Journal Royal Statistical Society B*, **64** (Part 2), 252–80.

Jorion, P. (1997) *Value at risk*. Irwin, Chicago–London–Singapore.

McNeil, A.J., Frey, R., and Embrechts, P. (2005) *Quantitative Risk Management*. Princeton University Press, Princeton and Oxford.

11 Validation of internal rating systems and PD estimates*

Dirk Tasche†

Abstract

Based on the theoretical binary classification framework, the notions of discriminatory power of a rating system and calibration of PD (probability of default) estimates are introduced. A variety of tools for measuring and testing discriminatory power and for testing correct calibration is presented.

1. Introduction

This chapter elaborates on the validation requirements for rating systems and probabilities of default (PDs) that were introduced with the New Capital Standards (commonly called 'Basel II', cf. BCBS, 2004). We start in Section 2 with some introductory remarks on the topics and approaches that will be discussed later on. Then, we have a view on the developments in banking regulation that have enforced the interest of the public in validation techniques. When doing so, we put the main emphasis on the issues with *quantitative validation*. The techniques discussed here could be used to meet the quantitative regulatory requirements. However, their appropriateness will depend on the specific conditions under which they are applied.

To have a common ground for the description of the different techniques, we introduce in Section 3 a theoretical framework that will be the basis for further considerations. Intuitively, a good rating system should show higher PDs for the less creditworthy rating grades. Therefore, in Section 4, we discuss how this monotonicity property is reflected in the theoretical framework from Section 3.

In Section 5, we study the meaning of *discriminatory power* and some tools for measuring it in some detail. We will see that there are tools that might be more appropriate than others for the purpose of regulatory validation of discriminatory power. The topic in Section 6 is *calibration of rating systems*. We introduce some of the tests that can be used for checking correct calibration and discuss the properties of the different tests. We then conclude in Section 7 with some comments on the question which tools might be most appropriate for quantitative validation of rating systems and PDs.

* The opinions expressed in this paper are those of the author and do not necessarily reflect views of Fitch Ratings.
† Fitch Ratings Ltd., London, United Kingdom.

2. Regulatory background

There is a long tradition of rating agencies grading firms that issue bonds. This aims primarily at facilitating the decision making of investors. Very roughly, the rating methodology applied by the agencies could be decribed as expert judgment that is based on hard as well as on soft facts.

Credit institutions have another 50 years long tradition of scoring borrowers. This way, the credit institutions want to support credit decisions, i.e. decisions to grant credit or not. With regard to scoring, the predominant methodology applied by the credit institutions could roughly be described as using statistically based score variables.

In the past, rating and scoring were regarded as being rather different concepts. This was partly caused by the fact that rating and scoring respectively are usually applied to populations with rather different characteristics. Ratings are most frequently used for pricing of bonds issued by larger corporates. Score variables are primarily used for retail credit granting.

But, also the background of the developers of rating methodologies and scoring methodologies, respectively, is usually quite different. Rating systems are often developed by experienced practitioners, whereas the development of score variables tends to be conferred on experts in statistics.

With the rising of modern credit risk management, a more unified view of rating and scoring has become common. This is related to the fact that today both rating and score systems are primarily used for determining PDs of borrowers. PDs are the crucial determinants for pricing and granting credit as well as for allocating regulatory and internal capital for credit risks.

In BCBS (2004), the Basel Committee on Banking Supervision recommends to take rating and scoring as the basis for determining risk-sensitive regulatory capital requirements for credit risks (Basel II). Compared to the Basel I standard, where capital requirements are uniformly at 8 per cent in particular for corporate borrowers irrespective of their creditworthiness, this is a major progress.

Credit institutions that apply the Basel II standardized approach can base the calculation of capital requirements on agency ratings, which are called *external ratings* in the Basel II wording. However, at least in continental Europe, external ratings are available only for a minority of the corporate borrowers. As a consequence, in practice, the capital requirements according to the standardized approach will not differ much from the requirements according to the Basel I regime.

Credit institutions that are allowed to apply the internal ratings-based (IRB) approach will have to derive PDs from ratings or scores they have determined themselves. Such ratings or scores are called *internal ratings*. The PDs then are the main determinants of the regulatory capital requirements. Note that in the IRB approach, capital requirements depend not only on PD estimates but also on estimates of loss given default (LGD) and exposure at default (EAD) parameters. Validation of LGD and EAD estimates is not a topic in this chapter.

As mentioned earlier, there are different ways to develop internal rating systems. On the one hand, there is the traditional approach to rating that is primarily based on expert knowledge. The number of rating grades is fixed in advance and assignments of grades are carried out according to qualitative descriptions of the grades in terms of economic strength and creditworthiness.

On the other hand, another – also more or less traditional approach – is scoring, which is primarily based on statistical methods. The first result then is a score variable that takes on values on a continuous scale or in a discrete range with many possible outcomes. The Basel II IRB approach requires that the score values are then mapped on a relatively small number of rating grades (at least seven non-default grades) but leaves the exact number of grades in the institution's discretion.

Combinations of rating systems that are based on statistical models and rating systems that are based on expert knowledge are called *hybrid* models. All kinds of combinations appear in practice, with quite different combination approaches. Driven partly by an IRB approach requirement, hybrid models even seem to be predominant. Often, they occur in the shape of a statistical model whose output can be overridden by expert decisions.

Among the rules on validation in the Basel II framework, two are particularly relevant for statistically based quantitative validation (see BCBS, 2004, §500):

1. Banks must have a robust system in place to validate the accuracy and consistency of rating systems, processes, and the *estimation of all relevant risk components*.
2. A bank must demonstrate to its supervisor that the internal validation process enables it to assess the *performance of internal rating* and risk estimation systems consistently and meaningfully.

The BCBS has established the Accord Implementation Group (AIG) as a body where supervisors exchange minds on implementation questions and provide general principles for the implementation of the Basel II framework. In particular, the AIG has proposed general principles for validation. Most of these principles are related to the validation process as such, and only some are relevant for quantitative validation. In the following list of principles (cf. BCBS, 2005a), the ones relevant for quantitative validation are emphasized.

(i) Validation is fundamentally about assessing the *predictive ability of a bank's risk estimates* and the use of ratings in credit processes.
(ii) The bank has primary responsibility for validation.
(iii) Validation is an iterative process.
(iv) There is *no single validation method*.
(v) Validation should encompass both *quantitative* and qualitative elements.
(vi) Validation processes and outcomes should be subject to independent review.

Hence, in particular, the Basel Committee emphasizes that validation is not only a quantitative statistical issue but also involves an important qualitative process-oriented component. This qualitative component of validation is commonly considered equally if not more important than the quantitative component. This chapter, however, deals with the quantitative component of validation only.[1]

Principle (1) of the AIG introduces the term 'predictive ability'. This is not a common statistical notion. It is not *a priori* clear whether it is related to well-known technical terms such as 'unbiasedness', 'consistency' and so on. However, there seems to be a consensus in the financial industry that 'predictive ability' should be understood in terms of *discriminatory power* and correctness of *calibration* of rating systems. We follow this path of interpretation for the rest of the chapter.

Commonly, discriminatory power is considered to be related to the discrimination between 'good' and 'bad' borrowers.[2] Additionally, there is a connotation of discriminatory power with the correctness of the ranking of the borrowers by the rating system. Whereas the importance of discriminatory power is obvious, examining the ranking seems to be of secondary importance, as in the end the ranking should be according to the size of the PD estimates. Therefore, correct ranking will be reached as soon as the calibration of the rating system is correct. This is a consequence of the fact that correct calibration is usually understood as having found the 'true' PDs for the rating grades. Correctness of the calibration of a rating system may be understood as implementation of the Basel Committee's requirement to assess the quality of the *estimation of all relevant risk components*. Checking discriminatory power may be interpreted as implementation of the Basel Committee's requirement to validate the *performance of internal rating*.

With regard to quantitative validation, the Basel Committee states in §501 of BCBS (2004) 'Banks must regularly *compare realised default rates with estimated PDs for each grade* and be able to demonstrate that the realised default rates are within the expected range for that grade'. Hence, there is a need for the institutions to compare PD estimates and realized default rates at the level of single rating grades. Such a procedure is commonly called *back-testing*.

In §502, the committee declares 'Banks must *also use other quantitative validation tools and comparisons with relevant external data sources*'. Thus, institutions are required to think about further validation methods besides back-testing at grade-level.

In §504 of BCBS (2004), the Basel Committee requires that 'Banks must have well articulated internal standards for situations where *deviations in realised PDs, LGDs and EADs from expectations become significant* enough to call the validity of the estimates into question. These standards must take account of *business cycles and similar systematic variability* in default experiences'. As a consequence, institutions have to decide whether perceived differences of estimates and realized values are really significant. Additionally, the committee expects that validation methods take account of systematic dependence in the data samples used for estimating the risk parameters PD, LGD and EAD.

In the retail exposure class, institutions need not apply fully fledged rating systems to determine PDs for their borrowers. Instead, they may assign borrowers to pools according to similar risk characteristics. Obviously, the requirements for quantitative validation introduced so far have to be modified accordingly.

3.   Statistical background

3.1.   Conceptual considerations

The goal with this section on the statistical background is to introduce the model that will serve as the unifying framework for most of the more technical considerations in the part on validation techniques. We begin with some conceptual considerations.

We look at rating systems in a *binary classification* framework. In particular, we will show that the binary classification concept is compatible with the idea of having more than two rating grades. For the purpose of this chapter, binary classification is understood in the sense of discriminating between the populations of defaulters and non-defaulters, respectively.

Also for the purpose of this chapter, we assume that the score or rating grade S (based on regression or other methods) assigned to a borrower summarizes the information that is contained in a set of covariates (e.g. accounting ratios). Rating or score variable design, development or implementation[3] is not the topic of this presentation. We want to judge with statistical methods whether rating or scoring systems are appropriate for discrimination between 'good' and 'bad' and are well-calibrated.

With regard to calibration, at the end of the section, we briefly discuss how PDs can be derived from the distributions of the scores in the population of the defaulters and non-defaulters, respectively.

3.2. Basic setting

We assume that with every borrower two random variables are associated. There is a variable S that may take on values across the whole spectrum of real numbers. And there is another variable Z that takes on the values D and N only. The variable S denotes a score on a continuous scale that the institution has assigned to the borrower. It thus reflects the institutions's assessment of the borrower's creditworthiness. We restrict our considerations to the case of continuous scores as this facilitates notation and reasoning. The case of scores or ratings with values in a discrete spectrum can be treated much in a similar way. The variable Z shows the state the borrower will have at the end of a fixed time-period, say after 1 year. This state can be *default*, D, or *non-default*, N. Of course, the borrower's state in a year is not known today. Therefore, Z is a *latent* variable.

The institutions's intention with the score variable S is to forecast the borrower's future state Z, by relying on the information on the borrower's creditworthiness that is summarized in S. In this sense, scoring and rating are related to binary classification. Technically speaking, scoring can be called binary classification with a one-dimensional co-variate.

3.3. Describing the joint distribution of (S,Z) with conditional densities

As we intend statistical inference on the connections between the score variable S and the default state variable Z, we need some information about the joint statistical distribution of S and Z. One way to describe this joint distribution is by specifying first the marginal distribution of Z and then the conditional distributions of S given values of Z.

Keep in mind that the population of borrowers we are considering here is composed by the sub-population of future defaulters, characterized by the value D of the state variable Z, and the sub-population of borrowers remaining solvent in the future, characterized by the value N of the state variable Z. Hence, borrowers with $Z = D$ belong to the defaulters' population and borrowers with $Z = N$ to the non-defaulters' population.

The marginal distribution of Z is very simple as it suffices to specify p, the *total probability of default* (also called *unconditional PD*) in the whole population. Hence, p is the probability that the state variable Z takes on the value D. It also equals 1 minus the probability that Z takes on the value N.

$$p = P[Z = D] = 1 - P[Z = N]. \tag{3.1a}$$

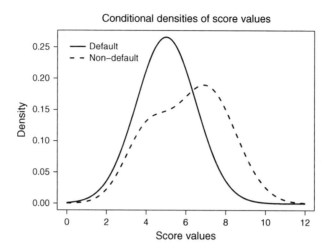

Figure 11.1 Illustrative example of score densities conditional on the borrower's status (default or non-default)

Note that the conditional probability of default given that the state variable takes on D is just 1 whereas the conditional PD given that the state is N is just 0. As the score variable S is continuous by assumption, its conditional distributions given the two values of Z can take on may be specified by *conditional densities* f_D and f_N, respectively. Figure 11.1 illustrates how a plot of such conditional densities might look like. The probability that the score S is not greater than some value s given that the state variable Z equals – say – D can be expressed as an integral of the density f_D.

$$F_z(s) = P[S \le s | Z = z] = \int_{-\infty}^{s} f_z(u)du, \quad z = D, N \tag{3.1b}$$

3.4. Describing the joint distribution of (S,Z) with conditional PDs

Another, in a sense, dual way of describing the joint distribution of S and Z is to specify for every value the score variable can take on the conditional probability $P[Z = D | S = s]$ that the state variable Z equals D. This is nothing but the conditional PD given the score. See Figure 11.2 for an example of how the graph of such a conditional PD function could look like.

$$\begin{aligned} s \mapsto P[Z = D | S = s] &= P[D | S = s] \\ &= 1 - P[N | S = s]. \end{aligned} \tag{3.2a}$$

To fully specify the joint distribution, in a second step then the unconditional distribution of the score variable S has to be fixed, e.g. via an unconditional density f.

$$P[S \le s] = \int_{-\infty}^{s} f(u)du. \tag{3.2b}$$

Note the difference between the unconditional density f of the score variable S on the one hand and the on the state variable Z-conditioned densities f_D and f_N on the

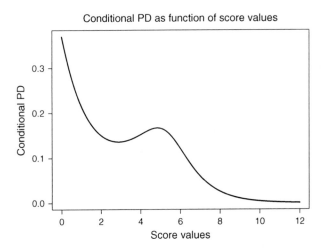

Figure 11.2 Illustrative example of probabilities of default (PDs) conditional on score values. Calculated with the densities from Figure 11.1 according to Equation 3.3b. Total PD 10 per cent

other hand. The unconditional density f gives the distribution of the scores in the whole population, whereas f_D and f_N describe the score distribution on sub-populations only. If the score variable S really bears information about the default state, then the three densities will indeed be different. If not, the densities might be similar or even identical. By means of the densities f, f_D and f_N, the distributions of all, the 'defaulting' and the 'non-defaulting', respectively, borrowers' score-variables are determined.

3.5. Equivalence of the both descriptions

So far, we have seen two descriptions of the joint distribution of the score variable and the state variable, which are quite different at first glance. However, thanks to Bayes' formula both descriptions are actually equivalent.

Suppose, first that a description of the joint distribution of score and state by the total probability of default p and the two conditional densities f_D and f_N according to Equations 3.1a and 3.1b is known. Then, the unconditional score density f can be expressed as

$$f(s) = p\, f_D(s) + (1-p)\, f_N(s), \tag{3.3a}$$

and the conditional PD given that the score variable takes on the value s can be written as

$$P[D|S = s] = \frac{p\, f_D(s)}{f(s)}. \tag{3.3b}$$

Assume now that the unconditional score density f in the sense of Equation 3.2b and the function representing the conditional PDs given the scores in the sense of Equation 3.2a are known. Then, the total PD p can be calculated as an integral of the unconditional density f and the conditional PD as

$$p = \int_{-\infty}^{\infty} P[D|S = s]\, f(s)\, ds, \tag{3.4a}$$

and both the conditional densities of the score variable can be obtained via

$$f_D(s) = P[D|S = s]\, f(s)/p \quad \text{and}$$
$$f_N(s) = P[N|S = s]\, f(s)/(1 - p). \tag{3.4b}$$

3.6. A comment on conditional PDs

It is useful to keep in mind that Equation 3.3b is only one of several ways to calculate conditional PDs. By definition, the conditional PD, $P[D|S]$, can also be described as the best forecast of the default/non-default state variable Z by a function of S in the least squares sense, i.e.

$$P[D|S] = \arg \min_{Y=f(S),\, f \text{ function}} E\left[(Z - Y)^2\right]. \tag{3.5}$$

This means that the conditional PD can be regarded as the solution of an optimization problem where the objective is to approximate as best as possible the state variable by some function of the score variable. Or, alternatively, it can be stated that given the information by the score S, there is no better approximation (in the least squares sense) of the state variable Z than $P[D|S]$. Intuitively, this is quite clear, because obviously a conditional PD of – say – 90 per cent would indicate that the borrower under consideration is close to default.

Note that because of the continuous distribution of S for any $s \in \mathbb{R}$ we have $P[S = s] = 0$. Hence, $P[D|S]$ is a conditional probability in the non-elementary sense and has to be dealt with carefully.

3.7. Dealing with cyclical effects

Recall from Section 2 that, for modelling, estimation and validation, institutions have to take account of 'business cycles and similar systematic variability'. We consider here how such cyclical effects, expressed as additional dependence on time, can be incorporated in the model framework we have introduced.

A first way to incorporate time-dependence is to assume that the conditional PDs, $P[D|S = \cdot]$, provided by the model are constant over time and to admit time-dependence only through the unconditional score density f, i.e. f is replaced by $f(\cdot, t)$. This corresponds to the so-called *through-the-cycle* (TTC) rating philosophy, where rating grades are assumed to express the same degree of creditworthiness at any time and economic downturns are only reflected by a shift of the score distribution towards the worse scores.

A second possibility to take account of time-dependence could be to assume that the conditional PDs are varying with time, i.e.

$$P[D|S = \cdot] = P_t[D|S = \cdot]. \tag{3.6}$$

This can be modelled with an assumption of constant conditional score densities f_D and f_N, having only the total PD $p = p_t$ time-dependent. Through Bayes' formula (3.3b), the conditional PDs would then depend upon time too. This approach corresponds to the *point-in-time* (PIT) rating philosophy according to which one and the same rating grade can reflect different degrees of creditworthiness, depending on the state of the economy.

3.8. The situation in practice

If we think in terms of a statistically based rating or scoring system, we can expect that from the development of the score variable conditional densities of the scores in the two populations of defaulters and non-defaulters respectively are known. In some institutions, it is then practice to predict the total probability of default for the following period of time by means of a regression on macro-economic data. In such a situation, Bayes' formula is useful for deriving the conditional PDs given the values of the score variable. Note that we are then in the PIT context we have described above. In some cases, for instance when logit or probit regression is applied, the score variable itself can be interpreted as conditional PD. This would again be a case of a PIT rating philosophy.

The popular software by Moody's-KMV provides a further example of this type. There, one can also say that the score variable is identical with the conditional probability of default. The traditional Moody's (or S&P or Fitch) ratings are commonly considered to be examples of the TTC rating philosophy.

3.9. Mapping score values on rating grades

Because of our assumptions and hence relevant for many score variables in practice, the theoretical probability that the score variable S takes on some fixed score value s is 0. As a consequence of this fact, PDs conditional on single scores s cannot directly be back-tested as there will not be many or even no observations of borrowers with score s. To facilitate validation, therefore, the BCBS decided to admit for the IRB approach only rating systems with a finite number of grades. Thus, if a rating system is based on a continuous score variable, a reasonable mapping of the scores on the grades must be constructed. In the remaining part of this section, we show how such a mapping can be constructed while taking into account the intended PDs of the grades.

The main issue with mapping score values on rating grades is to fix the criterion according to which the mapping is defined. We consider two different quantitative criteria. The first criterion for the mapping we consider is the requirement to have *constant PDs over time*.

To describe the details of the corresponding mapping exercise, assume that k non-default rating grades have to be defined. Grade 1 denotes the grade that indicates the highest creditworthiness. Increasing PDs $q_1 < \ldots < q_{k-1}$ have been fixed. Assume additionally that the conditional PD, $P[D|S = s]$, as a function of the score values is decreasing in its argument s. We will come back to a justification of this assumption in Section 4.

Given the model framework introduced above, the theoretical conditional PD given that the score variable takes on a value equal to or higher than a fixed limit s_1 can be determined. This observation also holds for the PD conditional on the event that the score

variable takes on a value between two fixed limits. Speaking technically, it is possible, b
proceeding recursively, to find limits $s_1 > s_2 > \ldots > s_{k-1}$ such that

$$q_1 = P[D|S \geq s_1] = \frac{p \int_{s_1}^{\infty} f_D(u)\, du}{\int_{s_1}^{\infty} f(u)\, du} \quad \text{and}$$

$$q_i = P[D|s_{i-1} > S \geq s_i] = \frac{p \int_{s_i}^{s_{i-1}} f_D(u)\, du}{\int_{s_i}^{s_{i-1}} f(u)\, du} \quad \text{for} \quad i = 2, \ldots, k-1.$$

(3.7a)

If a borrower has been assigned a score value s equal to or higher than the limit s_1, h
or she receives the best grade $R(s) = 1$. In general, if the score value is less than a limi
s_{i-1} but equal to or higher than the next limit s_i then the intermediate grade $R(s) = i$ i
assigned. Finally, if the borrower's score value is less than the lowest limit s_{k-1} then h
or she receives the worst grade $R(s) = k$.

$$R(s) = \begin{cases} 1 & \text{falls } s \geq s_1, \\ i & \text{falls } s_{i-1} > s \geq s_i, i = 2, \ldots, k-1, \\ k & \text{falls } s_{k-1} > s. \end{cases}$$

(3.7b)

The PD of the worst grade k cannot be fixed in advance. It turns out that its valu
is completely determined by the PDs of the better grades. This value, the conditiona
PD given that the rating grade is k, can be calculated in a way similar to the calcula
tions of the conditional PDs for the better grades in Equation 3.7a. As the result w
obtain

$$P[D|\, R(S) = k] = P[D|\, s_{k-1} > S] = p\, \frac{\int_{-\infty}^{s_{k-1}} f_D(u)\, du}{\int_{-\infty}^{s_{k-1}} f(u)\, du}.$$

(3.7c)

As $s \mapsto P[D|\, S = s]$ decreases, $r \mapsto P[D|\, R(S) = r]$ is an increasing function, as should b
expected intuitively.

The second criterion for the mapping we consider is the requirement to have a *constan
distribution of the good borrowers across the grades over time*. The intention with suc
a criterion might be to avoid major shifts with respect to ratings in the portfolio whe
the economy undergoes a downturn.

Assume, hence, that a number k of non-default grades (grade 1 best) and shares
$0 < r_1, \ldots, r_k, \sum_{i=1}^{k} r_i = 1$, of good borrowers in the grades have been fixed. Assume agai
additionally that high score values indicate high creditworthiness. It turns out that it i
then possible, again by proceeding recursively, to find limits $s_1 > s_2 > \ldots > s_{k-1}$ such tha

$$r_1 = P[S \geq s_1|N] = \int_{s_1}^{\infty} f_N(s)\, ds,$$

$$r_i = P[s_{i-1} > S \geq s_i|N] = \int_{s_i}^{s_{i-1}} f_N(s)\, ds, \quad i = 2, \ldots, k-1.$$

(3.8a)

The mapping of the scores on grades in this case is again defined by Equation 3.7b. A
$\sum_{i=1}^{k} r_i = 1$, this definition implies immediately

$$P[R(S) = k|N] = P[S < s_{k-1}|N] = r_k.$$

(3.8b)

Note that in both cases of mapping criteria we have considered, to keep constant over time the PDs and the shares of good borrowers respectively, the limits $s_1 \ldots, s_{k-1}$ must be periodically updated.

4. Monotonicity of conditional PDs

The first mapping procedure described in the last part of Section 3 works under the assumption that the conditional PD given the score $P[D|S = s]$ is a function that decreases in its argument s. As Figure 11.2 demonstrates this need not be the case. Are there any reasonable conditions such that monotonicity of the conditional PDs given the score is guaranteed?

An answer[4] to this question, in particular, will provide us with a justification of the monotonicity assumption which underlies Equations 3.7a–7c. This assumption is needed to ensure that the proposed mapping procedure for having constant PDs over time really works.

We discuss the question in the context of a *hypothetical decision problem*. Assume that we consider a borrower chosen at random and have been informed about his or her score value. But, of course, we do not yet know whether the borrower will default. How could we infer the value of his or her default state variable Z? In formal terms: suppose that a realization (s, z) of (S, Z) has been sampled. s is observed, z is not yet visible. Is $z = N$ or $z = D$?

One way to come to a decision would be to fix some set A of score values such that we infer default, D, as state if the borrower's score value is in A. If the value of the score were not in A, we would conclude that the state is N for non-default, i.e.

$$s \in A \Rightarrow \text{Conclusion } z = D$$
$$s \notin A \Rightarrow \text{Conclusion } z = N. \qquad (4.1)$$

How should we choose the *acceptance set* A? A convenient way to deal with the problem of how to find A is to have recourse to statistical test theory.

Then the problem can be stated as having to discriminate between the conditional score distributions $P[S \in \cdot | D]$ on the defaulters' and $P[S \in \cdot | N]$ on the non-defaulters' sub-populations, respectively. Thus, we have to decide whether the borrower under consideration stems from the defaulters' population or from the non-defaulters' population. The key concept for solving this problem is to have as objective a high certainty in the case of a decision to reject the presumption that the borrower is a future defaulter.

Formally, we can state this concept as follows: the null hypothesis is that the borrower is a future defaulter or, equivalently, that his or her state variable takes on the value D. The alternative hypothesis is that the borrower's state variable has got the value N.

- *Null hypothesis:* $P[S \in \cdot | D]$, i.e. $z = D$.
- *Alternative:* $P[S \in \cdot | N]$, i.e. $z = N$.

We conduct a statistical test on the null hypothesis 'state equals D' against the alternative 'state equals N'. Hence, our decision could be wrong in two ways. The so-called type I error would be to reject 'state equals D' although the state is actually D. The so-called type II error would be to accept 'state equals D' although 'state equals N' is true.

- *Type I error*: erroneously rejecting $z = D$.
- *Type II error*: erroneously accepting $z = D$.

To arrive at an optimal decision criterion, the probabilities of the two possible erroneous decisions have to be considered: The probability of the type I error is the probability under the defaulters' score distribution that a borrower's score will *not* be an element of the acceptance set A.

$$P[\text{Type I error}] = P[S \notin A|D]. \tag{4.2a}$$

In contrast, the probability of the type II error is the probability under the non-defaulters' score distribution that a borrower's score will be an element of the acceptance set A.

$$P[\text{Type II error}] = P[S \in A|N]. \tag{4.2b}$$

The type I error probability is usually limited from above by a small constant. Common values are 1 or 5 per cent, but we will see in Section 5 that for the purpose of validation also higher values make sense. Having bounded the type I error probability from above, the objective is to minimize the type II error probability.

$$\begin{aligned} &P[\text{Type I error}] \le (\text{small}) \text{ constant} \\ &P[\text{Type II error}] \text{ as small as possible.} \end{aligned} \tag{4.3}$$

Note that $1 - P[\text{Type II error}] = P[S \notin A|N]$ is called the *power of the test* we are conducting.

The optimal solution for the decision problem (given by Equation 4.3) is provided by the well-known Neyman–Pearson lemma. We state here briefly a slightly simplified version. See Casella and Berger (2001) or other textbooks on statistics for a more detailed version of the lemma. Let now α denote a fixed bound for the type I error probability, say α equal to 5 per cent. Hundred per cent minus such an α is called *confidence level* of the test.

The first step in stating the *Neyman–Pearson lemma* is to introduce another random variable, the so-called *likelihood ratio*. It is obtained by applying the ratio of the non-defaulters' and the defaulters' conditional score densities f_N and f_D, respectively, as a function to the score variable itself. The second step is to determine the $1 - \alpha$-quantile r_α of the likelihood ratio.

$$r_\alpha = \min\left\{ r \ge 0 : P\left[\frac{f_N}{f_D}(S) \le r \,|D\right] \ge 1 - \alpha \right\}. \tag{4.4a}$$

In the case of any reasonable assumption on the nature of the continuous conditional score distributions, this can be done by equating $1 - \alpha$ and the probability that the likelihood ratio is not higher than the quantile, i.e.

$$1 - \alpha = P\left[\frac{f_N}{f_D}(S) \le r_\alpha | D\right],\tag{4.4b}$$

and then solving the equation for the quantile r_α. Having found the quantile of the likelihood ratio, the decision rule 'Reject the hypothesis that the future state is D if the likelihood ratio is greater than the quantile' is optimal among all the decision rules that guarantee a type I error probability not greater than α. Hence, formally stated, the decision rule

$$\frac{f_N}{f_D}(S) > r_\alpha \quad \Leftrightarrow \quad \text{rejecting } D\tag{4.5a}$$

minimizes the type II error under the condition

$$P\,[\text{Type I error}] \le \alpha.\tag{4.5b}$$

As a consequence, for any decision rule of the shape

$$S \notin A \quad \Leftrightarrow \quad \text{rejecting } D\tag{4.6a}$$

and with $P[S \notin A|D] \le \alpha$ we have

$$P\left[\frac{f_N}{f_D}(S) \le r_\alpha | N\right] \le P\,[S \in A|N].\tag{4.6b}$$

In other words, any optimal test of D (defaulter) against N (non-defaulter) at level α looks like the likelihood ratio test, i.e. from Equation 4.6b follows

$$A = \left\{s : \frac{f_N(s)}{f_D(s)} \le r_\alpha\right\}.\tag{4.7}$$

Actually, Equation 4.7 is not only the optimal decision criterion for discriminating between defaulters and non-defaulters but also provides an answer to the original question of when the conditional PD given the scores is a monotonous function.

4.1. *Cut-off decision rules*

To explain this relation we need a further definition. A score variable S is called to be of *cut-off type* with respect to the distributions $P[S \in \cdot|D]$ and $P[S \in \cdot|N]$, if for every type I error probability α a decision rule of half-line shape

$$S > r_\alpha \quad \Leftrightarrow \quad \text{rejecting } D\tag{4.8a}$$

or for every α a rule of half-line shape

$$S < r_\alpha \quad \Leftrightarrow \quad \text{rejecting } D\tag{4.8b}$$

is *optimal* in the sense of minimizing the type II error probability under the constraint (Equation 4.5b). Decision rules as in Equations 4.8a and 4.8b are called *cut-off rules*.

By Equation 4.7, we can now conclude that

the score variable S is of cut-off type with respect to P[S ∈ ·|D] and P[S ∈ ·|N], if and only if the likelihood ratio s ↦ $f_N(s)/f_D(s)$ is monotonous.

Note that, for any score variable, its corresponding likelihood ratio is of cut-off type.

4.2. Conclusions for practical applications

Bayes' formula (Equation 3.3b) shows that the likelihood ratio is monotonous if and only if the conditional PD $s \mapsto P[D|S = s]$ is monotonous. There are some theoretical examples where the likelihood ratio is indeed monotonous: for instance, when both conditional densities f_N and f_D are normal densities, with equal standard deviation.

Unfortunately, in practice, monotonicity of the likelihood ratio or the conditional PD is hard to verify. However, from economic considerations, it can be clear that cut-off decision rules for detecting potential defaulters are optimal. This may justify the assumption of monotonicity. If, however, non-monotonicity of the likelihood ratio is visible from graphs as in Figure 11.2, the reliability of the score variable may be questioned. This yields a first example for a validation criterion for score variables, namely is the likelihood ratio monotonous or not?

5. Discriminatory power of rating systems

The following section is devoted to studying the question of how discriminatory power can be measured and tested. We have seen in Section 3 that the statistical properties of a score variable can to a high extent be expressed by the conditional densities of the score variable on the two populations of the defaulters and non-defaulters, respectively. Another, closely related way of characterization is by means of the conditional probability of default given the score values. With this observation in mind, discriminatory power can roughly be described in technical terms as discrepancy of the conditional densities, as variation of the conditional PD or as having the conditional PDs as close as possible to 100 per cent or 0 per cent.

Many statistical tools are available for measuring discriminatory power in one of these ways. We will consider a selection of tools that enjoy some popularity in the industry. A first major differentiation among the tools can be applied according to whether their use involves estimation of the total (or portfolio-wide or unconditional) probability of default. If this is necessary, the tool can be applied only to samples with the right proportion of defaulters. If estimation of the total PD is not involved, the tool can also be applied to non-representative samples. This may be important in particular when the power shall be estimated on the development sample of a rating system. The presentation of tools in the remaining part of this section closely follows the presentation in Chapter III of BCBS (2005b). Of course, the presented list of tools for measuring discriminatory power is not exhaustive. Which tool should be preferred may strongly depend on the intended application. As a consequence, various scientific disciplines such as statistics in medicine, signal theory or weather forecasting in the course of time suggested quite different approaches on how to measure the discriminatory power of classification systems.

5.1. Cumulative accuracy profile

The cumulative accuracy profile (CAP) is a useful graphical tool for investigating the discriminatory power of rating systems. Recall from Equation 3.1b the notions F_N and F_D for the distribution functions of the score variable on the non-defaulters' and defaulters', respectively, populations. Denoting by p as in Section 3 the total (portfolio-wide) probability of default, it follows from Equation 3.3a that the unconditional distribution function $F(s)$ of the score variable can be written as

$$F(s) = P\,[S \le s] = (1-p)F_N(s) + pF_D(s). \tag{5.1}$$

The equation of the CAP *function* is then given by

$$CAP\,(u) = F_D\left(F^{-1}(u)\right), \quad u \in (0, 1). \tag{5.2}$$

The graph of this function can either be drawn by plotting all the points $(u, CAP(u))$, $u \in (0, 1)$ or by plotting all the points $(F(s), F_D(s)), s \in \mathbb{R}$. The latter parametrization of the CAP curve can still be used when the score distribution function F is not invertible. Figure 11.3 shows examples of how a CAP curve may look like. The solid curve belongs to the score variable whose two conditional densities are shown in Figure 11.1. A curve like this could occur in practice. The two other curves in Figure 11.3 correspond to the so-called *random* (dotted line) and *perfect* (dashed curve), respectively, score variables. In case of a random score variable, the two conditional densities f_D and f_N are identical. Such

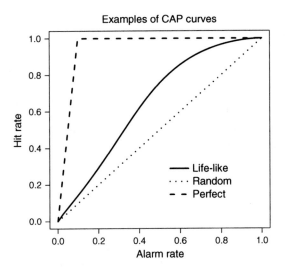

Figure 11.3 Illustrative example of CAP curves of life-like, random and perfect score variables as explained in main text. For the score variable associated with the conditional densities shown in Figure 11.1. Total PD 10 per cent. AR = 0.336

a score variable has no discriminatory power at all. In case of a perfect score variable, the densities f_D and f_N have disjoint supports, i.e.

$$\{s : f_D(s) > 0\} \cap \{s : f_N(s) > 0\} = \emptyset. \tag{5.3}$$

Equation 5.3 implies that the realizable range of the scores of the defaulting borrowers and the realizable range of the scores of the non-defaulting borrowers are disjoint, too. As a consequence, perfect discrimination of defaulters and non-defaulters would be possible.

In the context of CAP curves, $F(s)$ is called *alarm rate* associated with the score level s and $F_D(s)$ is called *hit rate* associated with the score level s. These notions indicate what happens if all the borrowers with a score equal to or less than some fixed threshold s are considered suspect of default (cut-off rule in the sense of Equations 4.8a and 4.8b). The hit rate then reflects which portion of the defaulters will be detected by this procedure. The alarm rate gives the portion of the whole population that will raise suspicion of being prone to default. From these observations, it follows that $100\,CAP(u)\%$ indicates the percentage of default-infected borrowers that are found among the first (according to their scores) $100u\%$ of all borrowers. A further consequence is that the 'perfect' curve in Figure 11.3 corresponds to a score variable that is of cut-off type in the sense of Section 4.

It seems unlikely that any rating system or score variable from practice will receive a CAP-curve like that from the perfect rating system as this would indicate that it will enable its owners to detect defaulters with certainty. Similarly, it is not very likely to observe in practice a rating system with zero power as a CAP-curve identical to the diagonal would indicate. However, if a rating system is developed for a certain portfolio and then is used on a completely different one, a very low discriminatory power can be the result of such a procedure.

It is easy to show that the CAP function is related to the conditional probability of default given the score through its derivative scaled by the total probability of default.

$$CAP'(u) = P[D|S = F^{-1}(u)]/p. \tag{5.4}$$

Recall from Equation 3.3b that by Bayes' formula the conditional probability of default given the score can be represented as a ratio involving the conditional score densities. As a consequence of Equation 5.4 and that representation, the stronger is the growth of $CAP(u)$ for u close to 0 (implying the conditional PD being close to 1 for low scores) and the weaker is the growth of $CAP(u)$ for u close to 1 (implying the conditional PD being close to 0 for high scores), the more differ the conditional densities and the better is the discriminatory power of the underlying score variable.

5.2. Accuracy ratio

From Figure 11.3, it is intuitively clear that the area between the diagonal line and the CAP curve can be considered a measure of discriminatory power. The random score variable receives area 0 and the life-like score variable obtains an area greater than 0 but less than the area of the perfect score variable. The area between the diagonal and the CAP-curve of the life-like rating system (solid line) can be calculated as the integral from 0 to 1 of the CAP function (Equation 5.2) minus 1/2. The area between the curve of the perfect score variable and the diagonal is given by $1/2 - p/2$ when p denotes the total probability of default.

The so-called *Accuracy Ratio* (AR) (also *Gini-coefficient*) is defined as the ratio of the area between the CAP-curve and the diagonal and the area between the perfect CAP curve and the diagonal, i.e.

$$AR = \frac{2 \int_0^1 CAP(u)du - 1}{1 - p}. \tag{5.5a}$$

Alternatively, the AR can be described as the difference of two probabilities. Imagine that two borrowers are independently selected at random, one from the defaulters' population and the other from the non-defaulters' population. The first probability is the probability of the event to observe a higher score for the non-defaulting borrower. The subtracted probability is the probability of the event that the defaulting borrower has the higher score. Then

$$AR = P[S_D < S_N] - P[S_D > S_N], \tag{5.5b}$$

where S_N and S_D are independent and distributed according to F_N and F_D, respectively. Obviously, if we assume that in general non-defaulters have the higher scores, we will expect that the first probability is higher than the second as is also indicated by the graph in Figure 11.3.

From Figure 11.3, we can conclude that the discriminatory power of a rating system will be the higher, the larger its AR is. This follows from Equation 5.4, as a large AR implies that the PDs for the low scores are large whereas the PDs for the high scores are small.

5.3. Receiver operating characteristic

The Receiver Operating Characteristic (ROC) is another graphical tool for investigating discriminatory power. Define, additionally to the notions of hit rate and alarm rate from the context of the CAP curve, the *false alarm rate* associated with the score level s as the conditional probability $P[S \le s|N] = F_N(s)$ that the score of a non-defaulting borrower is less than or equal to this score level. Then the false alarm rate reflects the portion of the non-defaulters' population, which will be under wrong suspicion when a cut-off rule with threshold s is applied. The equation of the ROC *function* is now given by

$$ROC(u) = F_D\left(F_N^{-1}(u)\right), \quad u \in (0, 1). \tag{5.6a}$$

The graph of this function can either be drawn by plotting all the points $(u, ROC(u))$, $u \in (0, 1)$ or by plotting all the points $(F_N(s), F_D(s))$, $s \in \mathbb{R}$. The latter parametrization of the ROC curve can still be used when the conditional score distribution function F_N is not invertible. In contrast to the case with CAP curves, constructing ROC curves does not involve estimation of the total (or portfolio-wide) probability of default. Figure 11.4 shows examples of how an ROC curve may look like. The solid curve belongs to the score variable whose two conditional densities are shown in Figure 11.1. The dotted and the dashed curves correspond to the random score variable and the perfect score

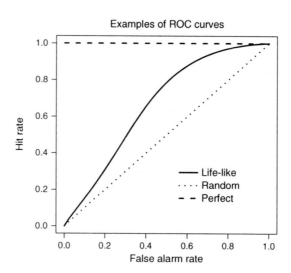

Figure 11.4 Illustrative example of Receiver Operating Characteristic ROC curves of life-like, random and perfect score variables as explained in main text. For the life-like score variable associated with the conditional densities shown in Figure 11.1. AUC = 0.668

variable respectively as in the case of the CAP curves in Figure 11.3. As a result of its definition, $100ROC(u)\%$ indicates the percentage of default-infected borrowers that have been assigned a score that is lower than the highest score of the first (according to their scores) $100u\%$ non-defaulters. Alternatively, the points on the ROC curve can be characterized as all pairs of type I error probability and power (see Section 4) that can arise when cut-off rules are applied for testing the hypothesis 'non-default' against the alternative 'default'.

The derivative of the ROC curve turns out to be closely related to the likelihood ratio that was already mentioned in the context of the Neyman–Pearson lemma in Section 4.

$$ROC'(u) = \frac{f_D(F_N^{-1}(u))}{f_N(F_N^{-1}(u))}, \qquad u \in (0, 1). \tag{5.6b}$$

Hence, the stronger is the growth of $ROC(u)$ for u close to 0 and the weaker is the growth of $ROC(u)$ for u close to 1, the more differ the conditional densities and the better is the discriminatory power of the underlying score variable.

From Section 4, we know that the score variable is of cut-off type and hence optimal in a test-theoretic sense if and only if the likelihood ratio is monotonous. Through Equation 5.6b, this is also equivalent to the ROC curve being concave or convex as concavity and convexity mean that the first derivative is monotonous. If high scores indicate high creditworthiness, the conditional score density f_D is small for high scores and large for low scores and the conditional score density f_N is large for high scores and small for low scores. As a conclusion, the ROC curve of an optimal score variable needs to be concave in the case where high scores indicate high creditworthiness. Although the lack of concavity of the solid curve in Figure 11.4 is not very clear, from the graph of its derivative according to Equation 5.6b in Figure 11.5 the lack of monotonicity is obvious.

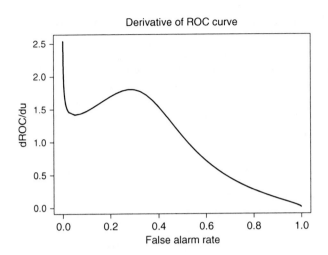

Figure 11.5 Derivative of the ROC curve given by the solid line in Figure 11.4. For the life-like score variable associated with the conditional densities shown in Figure 11.1

5.4. Area under the curve

As a further measure of discriminatory power, the *Area under the curve* (AUC) is defined as the area between the ROC curve and the axis of abscissa in Figure 11.4. This area can be calculated as the integral of the ROC curve from 0 to 1, i.e.

$$AUC = \int_0^1 ROC(u)du. \tag{5.7a}$$

Alternatively, the AUC can be described as a probability, namely that the score of a non-defaulter selected at random is higher than the score of an independently selected defaulting borrower.

Hence,

$$AUC = P[S_D < S_N], \tag{5.7b}$$

where S_N and S_D are independent and distributed according to F_N and F_D, respectively. Moreover, it can be proved (cf., for instance, Engelmann *et al.*, 2003) that the AUC is just an affine transformation of AR, namely

$$AUC = \frac{AR+1}{2}. \tag{5.7c}$$

As a consequence of this last observation, the higher the AUC the higher is the discriminatory power of the rating system under consideration, as is the case for AR. Moreover, maximizing AUC is equivalent to maximizing discriminatory power as maximizing the area under the ROC curve due (Equation 5.6b) and (Equation 3.3b) results in high PDs for small scores and low PDs for large scores. Additionally, Equation 5.7c shows that the value of AR – like the value of AUC – depends on the conditional densities of the score variable given the state of the borrower but not on the total probability of default in the portfolio.

There is also an important consequence from the representation of AUC as a probability. The non-parametric Mann–Whitney test (see, e.g. Sheskin, 1997) for the hypothesis that one distribution is stochastically greater than another can be applied as a test on whether there is discriminatory power at all. Additionally, a Mann–Whitney-like test for comparing the discriminatory power values of two or more rating systems is available (cf. Engelmann et al., 2003).

5.5. Error rates as measures of discriminatory power

We have seen that the ROC curve may be interpreted as a 'type I error level'-power diagram related to cut-off decision rules in the sense of Equations 4.8a and 4.8b, based on the score variable under consideration. Another approach to measuring discriminative power is to consider only total probabilities of error instead of type I and II error probabilities separately.

The first example of an error-rate-based measure of discriminatory power is the *Baysian error rate*. It is defined as the minimum total probability of error that can be reached when cut-off rules are applied.

$$
\begin{aligned}
\text{Baysian error rate} &= \min_s \text{P[Erroneous decision when cut-off rule with threshold} \\
&\quad s \text{ is applied]} \\
&= \min_s \left(P[Z = D] P[S > s | Z = D] + P[Z = N] P[S \le s | Z = N] \right) \quad (5.8a) \\
&= \min_s \left(p \left(1 - F_D(s) \right) + (1 - p) F_N(s) \right).
\end{aligned}
$$

In the special case of a hypothetical total PD of 50 per cent, the Baysian error rate is called *classification error*. Assume that defaulters tend to receive smaller scores than non-defaulters, or, technically speaking, that F_D is stochastically smaller than F_N [i.e. $F_D(s) \ge F_N(s)$ for all s]. The classification error can then be written as

$$
\text{Classification error} = 1/2 - 1/2 \max_s |F_D(s) - F_N(s)|. \quad (5.8b)
$$

The maximum term on the right-hand side of Equation 5.8b is just the population version of the well-known *Kolmogorov–Smirnov* statistic for testing whether the two distributions F_D and F_N are identical. The conditional distributions of the score variable being identical means that the score variable has not any discriminatory power. Thus, the classification error is another example of a measure of discriminatory power for which well-known and efficient test procedures are available. The so-called *Pietra-index* reflects the maximum distance of a ROC curve and the diagonal. In the case where the likelihood ratio f_D/f_N is a monotonous function, the Pietra-index can be written as an affine transformation of the Kolmogorov–Smirnov statistic and therefore is equivalent to it in a statistical sense.

If the likelihood ratio is monotonous, the Kolmogorov–Smirnov statistic has an alternative representation as follows:

$$
\max_s |F_D(s) - F_N(s)| = 1/2 \int_{-\infty}^{\infty} |f_D(s) - f_N(s)| \, ds \in [0, 1/2]. \quad (5.8c)
$$

This representation is interesting because it allows to compare the Kolmogorov–Smirnov statistic with the *information value*, a discrepancy measure that is based on relative entropies. We will not explain here in detail the meaning of relative entropy. What is important here is the fact that the information value can be written in a way that suggests to interpret the information value as something like a 'weighted Kolmogorov–Smirnov' statistic.

$$\text{Information value} = \text{E}\left[\log\frac{f_D(S)}{f_N(S)}\Big|D\right] + \text{E}\left[\log\frac{f_N(S)}{f_D(S)}\Big|N\right]$$
$$= \int_{-\infty}^{\infty}(f_D(s) - f_N(s))(\log f_D(s) - \log f_N(s))ds \tag{5.8d}$$
$$\in [0, \infty).$$

Note that the information value is also called *divergence* or *stability index*. Under the notion stability index, it is sometimes used as a tool to monitor the stability of score variables over time.

5.6. Measuring discriminatory power as variation of the PD conditional on the score

So far, we have considered measures of discriminatory power that are intended to express the discrepancy of the conditional distributions of the scores for the defaulters' population and the non-defaulters' population, respectively. Another philosophy of measuring discriminatory power is based on measuring the variation of the conditional PD given the scores. Let us first consider the two extreme cases.

A score variable has no discriminatory power at all if the two conditional densities of the score distribution (as illustrated in Figure 11.1) are identical. In that case, the borrowers' score variable S and state variable Z are stochastically independent. As a consequence, the conditional PD given the score is constant and equals the total PD.

$$P[D|S] = p. \tag{5.9a}$$

One could also say that the score variable S does not bear any information about potential default. Obviously, such a score variable would be considered worthless.

The other extreme case is the case where the conditional PD given the scores takes on the values 0 and 1 only.

$$P[D|S] = 1_D = \begin{cases} 1, & \text{if borrower defaults;} \\ 0, & \text{if borrower remains solvent.} \end{cases} \tag{5.9b}$$

This would be an indication of a perfect score variable, as in such a case there were no uncertainty about the borrowers' future state any more. In practice, none of these two extreme cases will occur. The conditional PD given the score will in general neither take on the values 0 and 1 nor be constant either.

In regression analysis, the determination coefficient R^2 measures the extent to which a set of explanatory variables can explain the variance of the variable which is to be predicted. A score variable or the grades of a rating system may be considered explanatory variables for the default state indicator. The conditional PD, given the score, is then the best predictor of the default indicator by the score in the sense of Equation 3.5. Its variance can be compared with the variance of the default indicator to obtain an R^2 for this special situation.

$$R^2 = \frac{\text{var}[P[D|S]]}{\text{var}[1_D]} = \frac{\text{var}[P[D|S]]}{p(1-p)} = 1 - \frac{E\left[(1_D - P[D|S])^2\right]}{p(1-p)} \in [0,1]. \tag{5.10a}$$

The closer the value of R^2 is to one, the better the score S can explain the variation of the default indicator. In other words, if R^2 is close to one, a high difference in the score values does more likely indicate a corresponding difference in the values of the default indicator variable. Obviously, maximizing R^2 is equivalent to maximizing $\text{var}[P[D|S]]$ and to minimizing $E[(1_D - P[D|S])^2]$.

The sum over all borrowers of the squared differences of the default indicators and the conditional PDs given the scores divided by the sample size is called *Brier score*.

$$\text{Brier score} = \frac{1}{n} \sum_{i=1}^{n} \left(1_{D_i} - P[D|S = S_i]\right)^2. \tag{5.10b}$$

The Brier score is a natural estimator of $E[(1_D - P[D|S])^2]$, which is needed for calculating the R^2 of the score variable under consideration. Note that as long as default or non-default of borrowers cannot be predicted with certainty [i.e. as long as Equation 5.9b is not satisfied] $E[(1_D - P[D|S])^2]$ will not equal 0.

In practice, the development of a rating system or score variable involves both an optimization procedure (such as maximizing R^2) and an estimation exercise (estimating the PDs given the scores $P[D|S = s]$). The Brier score can be used for both purposes. On the one hand, selecting an optimal score variable may be conducted by minimizing $E[(1_D - P[D|S])^2]$, which usually also involves estimating $P[D|S = s]$ for all realizable score values. On the other hand, when the score variable S has already been selected, the Brier score may be used for calibration purposes (see Section 6).

5.7. Further entropy measures of discriminatory power

Besides the information value defined in Equation 5.5 sometimes also other entropy-based measures of discriminatory power are used in practice.

For any event with probability p its *information entropy* is defined as

$$H(p) = -(p \log p + (1-p) \log(1-p)) \tag{5.11a}$$

Note from Figure 11.6 that $H(p)$ is close to 0 if and only if p is close to 0 or close to 1. As a consequence, information entropy can be regarded as a measure of uncertainty of the underlying event. When discriminatory power of a score variable has to be measured, it can be useful to consider the information entropy applied to the conditional PD given the scores, i.e. $H(P[D|S])$. If the average value of the information entropy is close to zero, the conditional PD given the scores will be close to zero or to one in average, indicating high discriminatory power. Formally, the average information entropy of the conditional

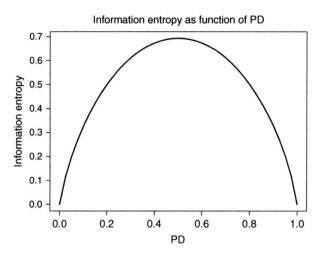

Figure 11.6 Graph of function $p \mapsto -(p \log p + (1-p)\log(1-p))$

PD is described as *conditional entropy* H_S, which is defined as the expectation of the information entropy applied to the conditional PD given the scores.

$$H_S = \mathrm{E}\left[H\left(\mathrm{P}[D|S]\right)\right]. \tag{5.11b}$$

As both the conditional PD given the scores as well as the calculation of the expectation depend on the portfolio-wide total PD, it is not sensible to compare directly the conditional entropy values of score variables from populations with different portions of defaulters. However, it can be shown by Jensen's inequality that the conditional entropy never exceeds the information entropy of the total probability of default of the population under consideration. Therefore, by using the *conditional information entropy ratio* (CIER), defined as ratio of information entropy of the total PD minus conditional entropy of the conditional PDs and the information entropy of the total PD, conditional entropy values of different score variables can be made commensurable.

$$\mathrm{CIER} = \frac{H(p) - H_S}{H(p)} \in [0, 1] \tag{5.11c}$$

The closer the value of CIER is to one, the more information about default the score variable S bears, in the sense of providing conditional PDs given the scores which are close to 0 or 1.

6. Calibration of rating systems

The issue with calibration of rating systems or score variables is how accurate the estimates of the conditional default probability given the score are. Supervisors, in particular, require that the estimates are not too low when they are used for determining regulatory capital requirements. In the following, we will consider some tests on calibration that

are conditional on the state of the economy. These are the binomial test, the Hosmer–
Lemeshow test and the Spiegelhalter test. As an example for unconditional tests, we will
then discuss a normal approximate test.

6.1. Conditional versus unconditional tests

The notions of conditional and unconditional tests in the context of validation for Basel
II can be best introduced by relating these notions to the notions of PIT and TTC PD
estimates (cf. Section 3.7 for the notions of PIT and TTC).

PD estimates can be based (or, technically speaking, conditioned) on the current state
of the economy, for instance by inclusion of macro-economic co-variates in a regression
process. The co-variates are typically the growth rate of the gross domestic product,
the unemployment rate or similar indices. The resulting PD estimates are then called
PIT. With such estimates, given an actual realization of the covariates, an assumption of
independence of credit events may be adequate, because most of their dependence might
have been captured by incorporating the economic state variables in the PDs estimates.

In contrast, unconditional PD estimates are not based on a current state of the economy.
Unconditional PDs that are estimated based on data from a complete economic cycle
are called TTC. When using unconditional PDs, no assumption of independence can be
made, because the variation of the observed default rates cannot be any longer explained
by the variation of conditional PDs, which are themselves random variables.

6.2. Binomial test

Consider one fixed rating grade specified by a range $s_0 \leq S \leq s_1$, as described, for instance,
in Equations 3.7a and 3.7b. It is then reasonable to assume that an average PD q has
been forecast for the rating grade under consideration. Let n be the number of borrowers
that have been assigned this grade.

If the score variable is able to reflect to some extent the current state of the economy,
default events among the borrowers may be considered stochastically independent. Under
such an independence assumption, the number of defaults in the rating grade is binomially
distributed with parameters n and q. Hence, the *binomial test* (cf., e.g. Brown *et al.*,
2001) may be applied to test the hypothesis 'the true PD of this grade is not greater than
the forecast q'. If the number of borrowers within the grade and the hypothetical PD q
is not too small, thanks to the central limit theorem under the hypothesis, the binomial
distribution can be approximated with a normal distribution. As already mentioned,
for this approximation to make sense is important that the independence assumption
is justified. This will certainly not be the case when the PDs are estimated TTC. The
following example illustrates what then may happen.

Example Assume that 1000 borrowers have been assigned the rating grade under consideration. The bank forecasts for this grade a PD of 1 per cent. One year after the forecast
19 defaults are observed.

If we assume independence of the default events, with a PD of 1 per cent the probability
to observe 19 or more defaults is 0.7 per cent. Hence, the hypothesis that the true PD is
not greater than 1 per cent can be rejected with 99 per cent confidence. As a consequence,
we would conclude that the bank's forecast was too optimistic.

Assume now that the default events are not independent. For the purpose of illustration, the dependence then can be modelled by means of a normal copula with uniform correlation 5 per cent (see, e.g. Pluto and Tasche, 2005, for details of the one-factor model). Then, with a PD of 1 per cent, the probability to observe 19 or more defaults is 11.1 per cent. Thus, the hypothesis that the true PD is not greater than 1 per cent cannot be rejected with 99 per cent confidence. As a consequence, we would accept the bank's forecast as adequate.

6.3.   Hosmer–Lemeshow test

The binomial test can be appropriate to check a single PD forecast. However, if – say – twenty PDs of rating grades are tested stand-alone, it is quite likely that at least one of the forecasts will be erroneously rejected. To have at least control over the probability of such erroneous rejections, joint tests for several grades have to be used.

So, assume that there are PD forecasts $q_1, \ldots, q_k$ for rating grades $1, \ldots, k$. Let n_i denote the number of borrowers with grade i and d_i denote the number of defaulted borrowers with grade i. The *Hosmer–Lemeshow statistic H* for such a sample is the sum of the squared differences of forecast and observed numbers of default, weighted by the inverses of the theoretical variances of the default numbers.

$$H = \sum_{i=1}^{k} \frac{(n_i q_i - d_i)^2}{n_i q_i (1 - q_i)}. \tag{6.1}$$

Under the usual assumptions on the appropriateness of normal approximation (like independence, enough large sample size), the Hosmer–Lemeshow statistic is χ^2_k-distributed under the hypothesis that all the PD forecasts match the true PDs. This fact can be used to determine the critical values for testing the hypothesis of having matched the true PDs. However, also for the Hosmer–Lemeshow test, the assumption of independence is crucial. Additionally, there may be an issue of bad approximation for rating grades with small numbers of borrowers.

6.4.   Spiegelhalter test

If the PDs of the borrowers are individually estimated, both the binomial test and the Hosmer–Lemeshow test require averaging the PDs of borrowers that have been assigned the same rating grade. This procedure can entail some bias in the calculation of the theoretical variance of the number of defaults. With the Spiegelhalter test, one avoids this problem.

As for the binomial and Hosmer–Lemeshow test, also for the Spiegelhalter test independence of the default events is assumed. As mentioned earlier, if the PD is estimated point in time, the independence assumption may be justified.

We consider borrowers $1, \ldots, n$ with scores s_i and PD estimates p_i. Given the scores, the borrowers are considered to default or remain solvent independently. Recall the notion of Brier score from Equation 5.10b. In contrast to the situation when a rating system or score variable is developed, for the purpose of validation we assume that realizations of the ratings are given and hence non-random. Therefore, we can drop the conditioning on the score realizations in the notation. In the context of validation, the Brier score is also called *Mean squared error (MSE)*.

$$MSE = 1/n \sum_{i=1}^{n} (1_{D_i} - p_i)^2, \tag{6.2a}$$

where 1_{D_i} denotes the default indicator as in (Equation 5.9b). The null hypothesis for the test is 'all PD forecasts match exactly the true conditional PDs given the scores', i.e. $p_i = P[D_i|S_i = s_i]$ for all i.

It can be shown that under the null we have

$$E[MSE] = 1/n \sum_{i=1}^{n} p_i(1 - p_i) \qquad \text{and} \tag{6.2b}$$

$$\text{var}[MSE] = n^{-2} \sum_{i=1}^{n} p_i(1 - p_i)(1 - 2p_i)^2. \tag{6.2c}$$

Under the assumption of independence given the score values, according to the central limit theorem, the distribution of the standardized MSE

$$Z = \frac{MSE - E[MSE]}{\sqrt{\text{var}[MSE]}} \tag{6.2d}$$

is approximately standard normally distributed under the null. Thus, a joint test of the hypothesis 'the calibration of the PDs with respect to the score variable is correct' can be conducted (see Rauhmeier and Scheule, 2005, for example from practice).

6.5. *Testing unconditional PDs*

As seen before by example, for unconditional PD estimates assuming independence of the defaults for testing the adequacy of the estimates could result in too conservative tests. However, if a time-series of default rates is available, assuming independence over time might be justifiable. Taking into account that unconditional PD estimates usually are constant[5] over time, a simple test can be constructed that does not involve any assumption of cross-sectional independence among the borrowers within a year. We consider a fixed rating grade with n_t borrowers (thereof d_t defaulters) in year $t = 1, \ldots, T$. Additionally, we assume that the estimate q of the PD common to the borrowers in the grade is of TTC type and constant over time, and that defaults in different years are independent. In particular, then the annual default rates d_t/n_t are realizations of independent random variables. The standard deviation σ of the default rates can in this case be estimated with the usual unbiased estimator

$$\hat{\sigma}^2 = \frac{1}{T-1} \sum_{t=1}^{T} \left(\frac{d_t}{n_t} - \frac{1}{T} \sum_{\tau=1}^{T} \frac{d_\tau}{n_\tau} \right)^2. \tag{6.3a}$$

If the number T of observations is not too small, and under the hypothesis that the true PD is not greater than q, the standardized average default rate is approximately standard normally distributed. As a consequence, the hypothesis should be rejected if the average default rate is greater than q plus a critical value derived by this approximation. Formally, reject 'true PD $\leq q$' at level α if

$$\frac{1}{T}\sum_{\tau=1}^{T}\frac{d_\tau}{n_\tau} > q + \frac{\hat{\sigma}}{\sqrt{T}}\Phi^{-1}(1-\alpha).$$ (6.3b)

As mentioned before, the main advantage of the *normal test* proposed here is that no assumption on cross-sectional independence is needed. Moreover, the test procedure seems even to be robust against violations of the assumption of inter-temporal independence, in the sense that the test results still appear reasonable when there is weak dependence over time. More critical appears the assumption that the number T of observations is large. In practice, time series of length 5–10 years do not seem to be uncommon. In Tables 11.1 and 11.2, we present the results of an illustrative Monte-Carlo simulation exercise to give an impression of the impact of having a rather short time series.

The exercise whose results are reflected in Tables 11.1 and 11.2 was conducted to check the quality of the normal approximation for the test of the unconditional PDs according to Equation 6.3b. For two different type I error probabilities, the tables present the true rejection rates of the hypothesis 'true PD not greater than 2 per cent' for different values of the true PDs. By construction of the test, the rejection rates ought to be not greater than the given error probabilities as long as the true PDs are not greater than 2 per cent. For the smaller error probability of 1 per cent this seems to be a problem, but not a serious one. However, the tables also reveal that the power of the test is rather moderate. Even if the true PD is so clearly greater than the forecast PD as in the case of 2.5 per cent, the rejection rates are only 19.6 and 30.1 per cent, respectively.

Table 11.1 Estimated PD $= 2\%$, T $= 5$, $\alpha = 1\%$

True PD(%)	Rejection rate(%)
1.0	0.00
1.5	0.01
2.0	2.05
2.5	19.6
5.0	99.2

Table 11.2 Estimated PD $= 2\%$, T $= 5$, $\alpha = 10\%$

True PD(%)	Rejection rate(%)
1.0	0.00
1.5	0.60
2.0	7.96
2.5	30.1
5.0	99.2

7. Conclusions

With regard to measuring discriminatory power, the AR and the Area under the Curve seem promising[6] tools as their statistical properties are well investigated and they are available together with many auxiliary features in most of the more popular statistical software packages.

With regard to testing calibration, for conditional PD estimates powerful tests such as the binomial, the Hosmer–Lemeshow and the Spiegelhalter test are available. However, their appropriateness strongly depends on an independence assumption that needs to be justified on a case-by-case basis. Such independence assumptions can at least partly be avoided, but at the price of losing power as illustrated with a test procedure based on a normal approximation.

References

Basel Committee on Banking Supervision (BCBS) (2004) Basel II: International Convergence of Capital Measurement and Capital Standards: a Revised Framework (http://www.bis.org/publ/bcbs107.htm).

Basel Committee on Banking Supervision (BCBS) (2005a) Update on work of the Accord Implementation Group related to validation under the Basel II Framework (http://www.bis.org/publ/bcbs_nl4.htm).

Basel Committee on Banking Supervision (BCBS) (2005b) Studies on the Validation of Internal Rating Systems (revised), Working Paper No. 14. (http://www.bis.org/publ/bcbs_wp14.htm).

Blochwitz, S., Hohl, S., Wehn, C. and Tasche, D. (2004) Validating Default Probabilities on Short Time Series *Capital & Market Risk Insights, Federal Reserve Bank of Chicago*.

Brown, L., Cai, T. and Dasgupta, A. (2001) Interval estimation for a binomial proportion. *Statistical Science* 16, 101–33.

Casella, G. and Berger, R.L. (2001) *Statistical inference*. Second edition. Duxberry, Pacific Grove.

Committee of European Banking Supervisors (CEBS) (2005) Guidelines on the implementation, validation and assessment of Advanced Measurement (AMA) and Internal Ratings Based (IRB) Approaches (http://www.c-ebs.org/pdfs/CP10rev.pdf).

Engelmann, B., Hayden, E. and Tasche, D. (2003) Testing rating accuracy. *Risk* 16, 82–6.

Pluto, K. and Tasche, D. (2005) Thinking positively. *Risk*, 18, 72–8.

Rauhmeier, R. and Scheule, H. (2005) Rating properties and their implications for Basel II capital. *Risk* 18, 78–81.

Sheskin, D.J. (1997) *Handbook of parametric and nonparametric statistical procedures*. CRC Press, Boca Raton.

Tasche, D. (2002) Remarks on the monotonicity of default probabilities, Working Paper (http://arxiv.org/abs/cond-mat/0207555).

Notes

1. For more information on qualitative validation see, e.g., CEBS (2005).
2. In this respect, we follow BCBS (2005b).
3. The process of design and implementation should be subject to qualitative validation.
4. This section is based on Tasche (2002).
5. Blochwitz *et al.* (2004) provide a modification of the test for the case of non-constant PD estimates.
6. The selection of the topics and the point of view taken in this chapter is primarily a regulatory one. This is caused by the author's background in a regulatory authority. However, the presentation does not reflect any official regulatory thinking. The regulatory bias should be kept in mind when the following conclusions are read. A procedure that may be valuable for regulatory purposes need not necessarily also be appropriate for bank-internal applications.

Index

CPSIA information can be obtained at www.ICGtesting.com
Printed in the USA
LVOW070844140613

338534LV00003B/9/P